GUIDING LIGHTS

A Tribute to Earth's Lighthouses

PAULINE BAIRD JONES

Perilously Fun Fiction

ISBN: 978-1-962125-76-5

Author Note

Like most things I do, the path to this tribute to lighthouses is a strange one. I've always loved lighthouses, despite being raise land-locked—or perhaps because of it?

Whatever the reason, these sentinels of safety have always fascinated me. I wrote one novel set in and around a lighthouse called *Specters in the Storm.* It was a lot of fun to research and write, but it didn't wholly satisfy my itch to write more stories in and around lighthouses.

I explored the idea of a lighthouse in space and produced *ScytheQ,* the sixth book in my Cyborg Chronicles series. Disasters in space are different from those on a planet, so of necessity, my space lighthouse is very different from these earth bound ones.

Most maritime disasters happen near the coast, not in the open sea as you would think. Back through time out of mind, sailors have paid with their lives when their vessels glided too close to the shore. A collision with rock, shoals, or sandbars is usually fatal.

Lighthouses and their keepers are the heroes, who save mariners from such a fate. The beacons guide them safely into harbors for refuge, commerce, or war, and when the ships wreck, the keepers risk their own lives to save those aboard.

These extraordinary landmarks, hold tales within them of war, lives saved, monstrous storms, and ghostly accounts of spirits, lingering within their walls. In the pages of this booklet are some of those cool and memorable lighthouse stories.

Because everyone loves lighthouses. There is something both romantic and bold about them standing sometimes on a little more than a rock in the middle of the vast ocean isolated daring and steadfast.

So in a weird way, this tribute is a companion piece to both books and to my life long love affair with lighthouses.

Introduction

Before they inspired the futuristic lighthouses in *ScytheQ: The Cyborg Chronicles 5*, and were the setting for my historical suspense, *Specters in the Storm,* real-world lighthouses served as beacons of hope, safety, and resilience across Earth's oceans. This booklet explores the fascinating history and enduring mystique of these coastal guardians. Whether you're a lover of maritime history or a fan of cyborgs and space adventures, discover how the light of the past illuminates the stories of the future.

Lighthouses in Antiquity

THE PHAROS LIGHTHOUSE, In The Harbor Of Alexandria, Egypt

"Its flame a second sun to guide seafarers," is how the Greek novelist, Achilles Tatius described the first stone lighthouse, the Pharos.

Ancient travelers took ships to the isle of Pharos to see the first stone lighthouse, built sometime between 297 and 283 BC on a massive square base that rose in tiers of white marble or limestone to an open cupola and on top of it stood a statue of Zeus, watching the seas, guarding Alexandria. For those were the days when men believed the gods walked among them.

Beneath the king of the gods, an open fire, magnified by polished bronze mirrors, lit up the sky, casting golden light far across dark, Mediterranean waters, and the billowing smoke of the eternally burning flames rose to the glistening stars.

A beacon guiding ships of old away from treacherous reefs looming beneath the crashing white foamed waves to the serene port of Alexandria.

Not only did it set the standards for beacon towers, and spur Romans, Spaniards, French, Italians and Turks, and other people to build lighthouses along their coast, it had another lasting impact on maritime history. In many languages, the word for lighthouse is a variant of Pharos. Faro in Spanish, Phare in French and the scientific study of lighthouses is Pharology.

Wouldn't you have liked to have been an ancient traveler 2320 years ago, hearing the roar of the waves lap against the island of Pharos, as you breathe in the tang of the sea, tasting its salty mist on your tongue?

You tilt your neck back to peer up 330, possibly 450, feet high to look at Zeus, on top. Dizziness assaults you. And the mid-day sun, glinting off white marble or limestone walls, almost blinds you. Your breath catches in your throat. You rub the goosebumps on your arms, and stand frozen in awe before the majesty of Pharos.

As powerful as Pharos was, it could not match Mother Nature's and Father Time's might. A series of earthquakes hit the area between 796 and 1323 AD, leaving this wonder of the world in ruins. From its stones, Sultan Qaitbay erected a medieval fort, the Citadel of Qaitbay, in 1480 AD.

THE TOWER OF HERCULES, La Coruna, Spain

The Tower of Hercules is the world's oldest and only working lighthouse from the ancient Roman empire. Built during Emperor Trajan's reign, 2nd century CE, and dedicated to the war god Mars. Imagine standing there when its fire was first lit. Watching history unfold as you discussed with those around of Hercules visit to this spot, after he slew Geryon. And how he buried that giant's head beneath the base of the light tower.

A reference in Historiae adversum Paganos, written by Paulus Orosius in 415–417 CE, has survived. "*At the second angle of the circuit circumnavigating Hispania, where the Gallican city of Brigantia is sited, a very tall lighthouse is erected among a few commemorative works, for looking towards Britannia."*

The fire burned out in the Middle Ages, when men no longer believed the gods lived among us. The Tower of Hercules fell into abandonment. But in the 17th century it was revived through renovation and later, completely rebuilt during the reign of Charles III.

Having stood nearly 2,000 years, its light still flashes every 20 seconds, guiding ships through the 'Coast of Death'. When you visit, after you admire the art of the tower, peruse the stunning paintings by major artists in the museum that is there now.

The Lighthouse of Ostia, Island of Portus, near Rome, Italy The bustling port at Ostia couldn't handle all the shiploads of grain and goods needed by the citizens of Rome. So, Emperor Claudius ordered construction on Portus, a massive harbor north of the city. The lighthouse of Ostia stood at the harbor's entrance, on an artificial island built on the sunken hull of a ship that once carried a great obelisk from Egypt to Rome.

Ancient mosaics and reliefs depict its elegant but practical arched openings and slit windows that allowed the light, fueled by burning wood or oil, to glare across the waters, visible to sailors braving the tides of the Tiber delta and treacherous sandbars.

Imagine during the days when the Roman empire ruled the world, standing at the lighthouse's base and admiring the flame that lit the way for the ships that fed the empire and kept it thriving.

The foundation of this ancient wonder is buried under a junkyard near Fiumicino. Its stones have crumbled, but we remember its glory days when Rome was the master of the seas.

Dark Age Lighthouses

. . .

Hooks Point Lighthouse, County Wexford, Ireland

Hooks Point La Coruna's Tower of Hercules remained in use throughout the medieval period, but with the fall of the Roman Empire, many lighthouses fell into disuse. And not much rebuilding occurred in the dark ages, though they used makeshift beacons.

For example, in the 5th century a Welsh monk, Dubhan, founded a monastery in County Wexford Ireland and set up a pile of rocks and a fire at Hooks Point—the tip of the peninsula at Waterford Harbour which opens into the Celtic Sea. The monks maintained the beacon, which guided ships through rough waters.

Years later, one of two men built a tower there.

You can go with the story that between 1170 and 1182, a Norman named Raymond LeGros landed at Baginbun in Wexford. Upon the remains of a Celtic fort, he built a Norman one where he battled and beat the Irish advance from Waterford. Before LeGros was slain in that battle, he erected the first Tower of Hook, around 1172.

The other man credited with the tower is the renown Norman knight, William Marshal, Earl of Pembroke. Many say in 1245, he built the stone tower which stood over 100 feet tall and 40 feet wide.

The lighthouse fell into disrepair, but Britain re-established it in 1667. Queen Anne transferred the administration of the lighthouse to Revenue Commissioners in 1704. But they didn't take over for a long time. Henry Loftus had a 21-year lease on it, then it passed to Henry's son. He threatened in 1726 to extinguish the light unless the Revenue Commissioners paid rent in advance. Finally, in 1791, the Revenue Commission granted a contract to maintain and staff Hook Point and several other Irish lighthouses.

Medieval Lighthouses

Scotland, Ireland, and Britain Lighthouses

Moving into the maritime medieval period, you're probably thinking Vikings or the Norman invasion of England.

• During that period, in medieval Ireland, nuns at Youghal used torches to guide vessels into the harbor. Then a lighthouse was built there in 1190 and was operated by the nuns until 1542 when the holy sisters left.

• A coal burning, English lighthouse, with two masonry towers used as range lights for navigating the River Tyne, was built at the mouth before 1600, then rebuilt in 1808.

• A small lighthouse at Saint Catherine's on the Isle of Wight was constructed in 1323 to help ships navigate the dangerous shoals along that coast.

• A hermit petitioned the king of England to support him in building a lighthouse on the eastern coast at Spurn Point to guide ships into the mighty Humber River. The king gave him permission to collect tolls for 10 years.

• Several lighthouses were constructed in Scotland during the medieval period. In 1556 Aberdeen, a portal lantern placed on a chapel guided ships into the harbor.

The Lighthouse of Genoa, Genoa Italy

the lighthouse, also known as the Lanterna, was built in the late medieval period, either 1128 or 1161. It's the world's fourth oldest, still operating, lighthouse.

In the early renaissance period, in 1318 and again in 1321, a trench was dug around the tower, to protect it from battle damage. To ensure incoming ships could see the signal fire better, a lantern fueled by olive oil was added in 1326, and the city's coat of arms was painted on it in 1340. Around 1400, the lighthouse was converted into a prison and one of the hostages held there was the king of Cyprus, James II, and his wife. In 1405, the priests who maintained the lighthouse placed Christian symbols of a fish and a golden cross on the cupola. Christopher Columbus's uncle, Antonio Colombo, was one of the lighthouse keepers in 1449. Then in 1543, the lighthouse was reconstructed into its current shape.

SHELLING by the French in 1684 damaged the windows, which Louis XIV replaced in 1692. In 1778, a new lighting system was added and in 1840 a rotating Fresnel lens was installed. Modifications continued until the end of the 19th century, and it was modernized again in 1913, but the electrification had to be refitted in 1936. After American and British air attacks in WWII, a major restoration was completed in 1956. To this day, it is one of the tallest lighthouses in the world, and a symbol of Genoa.

Renaissance Lighthouses

Now, from the 14th to 17th century, we enter the renaissance, meaning rebirth. An age of renewed interest in learning, creativity, and innovation. Per maritime history, you're probably thinking about the discovery of the new world and colonization.

The Kopu Lighthouse, Mägipe, Estonia

Kõpu is the third oldest lighthouse still standing. It was built in 1531 atop Hiiumaa's rugged cliffs in Estonia, overlooking the vast Baltic Sea. Visible for up to 12 miles offshore, it's guided mariners, on east-west trade routes, over treacherous shoals for nearly 500 years. First modernized in 1649, with an open iron fire grate on top, then firewood replaced coal to save on transportation costs. A change that nearly deforested the peninsula.

Mid-17th century, under Swedish rule, height was added, so it rose a full 117 feet and was enhanced with iron staircases. Later, under the care of the Russian Empire, oil lamps, reflectors, and, eventually, a clockwork mechanized rotating lantern was added. WWII, German bombers damaged the lantern, but the resilient tower stood firm. Post-war repairs brought electricity and a rotating system powered by magnetic fields. In 2020, it was enhanced with the world's most powerful LED light source, visible for 26 nautical miles.

Wouldn't you love to visit Kõpu. To stand in the lighthouse's shadow as the salty Baltic wind blows in your face, while you bear witness to the unwavering light that has guided sailors for five centuries.

Cordouan Lighthouse, France

At the edge of France's Atlantic coast, where the Gironde Estuary meets the open sea, the Cordouan Lighthouse stands above the waves.

Commissioned by King Henry III as a show of France's maritime power and architectural prowess, the brilliant architect, Louis de Foix, blended functionality with beauty.

In renaissance grandeur, pilasters, columns, modillions, and gargoyles adorn it. The interior holds a chapel decorated with intricate stonework, symbolizing the spiritual connection between humanity and the sea.

Erected between 1584 and 1611, it towers 221 feet. Cordouan is still a working lighthouse. Whether fueled by coal, oil, or modern electricity, the lantern that crowns it has guided sailors across treacherous waters since the early renaissance.

Eddystone Lighthouse, Boston

The first colonial beacon in America was erected September 14th, 1716, to guide ships into Boston Harbor. The Eddystone Lighthouse's 15-foot tower was crafted from rough-cut granite to withstand New England storms.

Perched atop it is an octagonal wooden lantern room, enclosed by glass, that housed the light from candles or crude whale oil lamps to guide mariners bringing goods, people, and ideas to the 13 colonies.

CONCLUSION

The Romans and Greeks built monumental towers woven into their myths and legends and laid the foundations for medieval beacons like Hook Head and Genoa's Lanterna. Then ornate lighthouses like the Kōpu and the Cordouan exemplified the spirit of the Renaissance age. We leave the Renaissance and enter the 18th century with the example of the Eddystone Lighthouse in the American British Colony of Massachusetts.

Now let's look at some of the valiant lighthouse keepers and the people they saved.

Heroic Lighthouse Keepers

HERMAN MELVILLE in Moby-Dick called lighthouse keepers "silent watchers."

IT WASN'T a rule that lighthouse keepers be married. But to the board and service, married men were less likely to drink too much or wander off. Plus, family men came with wives, and often children—a ready-made team of assistants the service or board didn't have to pay for.

PART I: REMARKABLE LIGHTHOUSE KEEPERS: MEN

MARCUS HANNA- CAPE ELIZABETH'S Lighthouse

Beneath the stinging, 10 below temperature of a Maine winter morning in 1885, Louise Hanna, a wife of head keeper of the lighthouse at Cape Elizabeth, opened the door to the keeper's house.

Her legs went as weak as noodles and her heart hammered, gazing at the haunting sight before her—the mast of a ship, tilted above the jagged rocks.

It was the Australian, its main sail shredded, and its deck coated in thick ice. All night long, Captain J.W. Lewis, and the other two men in his crew, battled whipping frost-laced gales and towering waves driving them at a wrecking speed to shore. Then, the Captain made the only decision he could, he wrecked the cargo ship onto the rocks, to save his crew.

Louise rushed back inside and alerted her husband, Marcus Hanna.

He trudged through frosty, waist-deep snowdrifts to the signal house, where his assistant, Hiram Staples, joined him. They hurried to the ocean's edge. Irving Pierce and William Keller's stiff, numb, nearly frozen fingers clung to the rigging. The men cried out over the thunderous, crashing waves, pleading for help. Captain Lewis was not there. Massive waves had engulfed him, plummeting him into the ocean's abyss.

Hanna could save them if he could get to the pilot house where his good rope was. He grabbed an axe and slogged through the relentless snow to the pilot house. Snow drifts barricaded the door. He went back for Staples who followed with a shovel, digging a path to the Piolet house, then Hanna chopped through the door with the axe, and snatched the rope.

Staples and Hanna returned to the roaring surf. With a piece of metal tied to the end of the rope, Hanna pitched it toward the freezing, near dead men, again and again. It fell short every time.

Suddenly, a massive wave slammed the Australian closer to shore, crashing it upon its beam ends. Exhausted, and nearly frozen, the two determined seaman still clung to the rigging and survival.

Their persistence and Hanna's paid off. The rope reached Keller, who tied it around Pierce's waist. Hanna hauled Pierce through the icy, heaving waves. His actual historical dialogue as he recounted that day was "Pierce's jaws were set, he was totally blind from exposure to the cold, and the expression on his face I shall not soon forget."

He tossed the rope again. Keller caught it and tied it around his waist. Finally, he'd be rescued.

But some time back, Hiram Stapples returned to the signal house. There was no one to help Hanna as numbing cold settled into his bones. Trembling uncontrollably. His mind groggy. Engulfed by weakness. Hardly able to move. How could he possibly pull this man to shore?

At that moment, Nathaniel Staples and two neighbors arrived. Giddiness spread through Hanna, and he exhaled in relief. The three men joined him and helped pull Keller ashore. Minutes later a towering wave smashed the Australia into nothing more than twigs littering the coast.

Hana, Nathaniel and the neighbors carried the two sailors to the signal house. There they stripped the ice coated clothes off them, dressed them in warm flannel, and revived them with hot drinks and food. Hanna and his assistants nursed them until the snowstorm died down enough for Hanna to ride to Portland and fetch doctors from the US marine hospital. Under their care, Pierce and Keller recovered from severe frostbite and shock.

For that extraordinary heroism, Marcus Hanna was awarded the Lifesaving Medal on April 29, 1885. He'd already earned the Congressional Medal of Honor during the Civil War, making him the only person in history awarded both medals, the military and the civilian equivalent, for going beyond what is expected, endangering his life to rescue others.

Harry C. Claiborne - Bolivar Point Lighthouse

The Bolivar Point lighthouse, built in 1872, after the first one was destroyed in the Civil War, stood 117 feet tall strong with a cast-iron shell and a brick-lined interior that the deadliest hurricane in U.S. history would soon test.

On September 8, 1900, Harry C. Claiborne, Head Keeper of the Bolivar Point Lighthouse in Galveston County, Texas, swung the iron door open, and let 100 Galvestonians in to shelter from the storm nearing land fall. The crowd settled on the spiral staircase, backs pressed against the cool iron, as howling winds rushed the lighthouse at 130 mph, with 200 mph gusts lashing everything in the hurricane's way.

Meanwhile, a quarter mile away, a train from Beaumont chugged over flooded tracks toward the Charlotte M Allen ferry to carry passengers across the Galveston ship channel to town. The conductor fought the swell of water engulfing the huge wheels by ordering the engineer to reverse the train. It didn't help. Instead, rising water flooded the coal compartments, bringing the train to a momentary stop. Ten passaging's grabbed the opportunity to take their chances on the lighthouse. They leapt off the train into the swirling, waist-high water to splash and wade their way to Bolivar Point, where Claiborne pulled them into his safe haven.

One hundred and twenty five people braced themselves as the gale battered the lighthouse, which groaned as it tilted side to side, Claiborne gripped the iron handles in the lantern room to keep from being thrown off his feet.

Claiborne wisely stocked up on food before the storm, but drinking water was another matter. Volunteers leaned over the catwalk with buckets in hand, catching the rain. The waves were so high that the first bucketfuls were salty. But persistence paid off, and they soon had fresh rainwater to quench their thirst.

By the next morning, the storm died down and moved on, so Claiborne opened the door. The scene haunted everyone. Warm sunshine bathed a landscape of destruction. The dead bodies of a dozen storm victims lay scattered at the base of the lighthouse. Galveston was basically leveled. Homes and buildings gone. At least 6,000 lives lost, some estimates reached 10,000. Among the dead were the 85 passengers who stayed on the train. The hurricane's rains engulfed the train and all those on it as if swallowing it whole.

Misjudged by the U.S. Weather Bureau as what we would now call a mild category 1, the storm, known as the 1900 Galveston Hurricane, caught everyone off guard. It struck with the full fury of a category four. To this day, the Galveston hurricane of 1900 remains the deadliest natural disaster in U.S. history.

Fifteen years later, August 16, 1915, another hurricane hit Galveston. Sixty Galvestonians rushed to the Bolivar Point Lighthouse, where Claiborne ushered them in and shut the door. That night, fierce, howling winds of 93 miles per hour, gusting up to 120 miles per hour, rocked the tower and broke the rotating lens' mechanism. Assistant Keeper James P. Brooks turned the lens manually until the lashing winds and water made it impossible. Brooks secured the lens and left the light burning. Then someone opened the door. Rolling water rushed in. Brooks plunged into the five feet deep floodwater, struggling to stay upright, he pressed on, lunging toward the door, slamming it shut.

The hurricane's savage winds kept them in another night. The storm extinguished the Bolivar Point beam. For the first time since it was built, the lighthouse stood dark. On August 18, 1915, the refugees emerged from the lighthouse to piles of rubble and debris where homes and buildings once stood.

Powerful gales, flooding of 6 to 9 feet with storm surges over 16 feet, left a direct death toll of 275, but possibly as high as 400, counting all the missing people. But the sixty people who sheltered in the lighthouse were alive and well, thanks to Claiborne, Brooks, and the Bolivar point lighthouse.

Mingo and the Harwich Lighthouses, Harwich England

As far back as 1667, possibly 1665, a Black man performed keeper duties at two lighthouses to help ensure mariners' safety into the Harwich port. Mingo was the first or one of the first lightkeepers of African descent in England. He stands out in maritime history as the first Black lighthouse keeper recorded by name. Sir William Batten, the Royal Navy Surveyor, and Master of the Trinity House was granted a patent for two lighthouses he erected at the port of Harwich. One on top of the Old Town Gate, its coal hearth casting a fiery glow for mariners. The light for the second one was a humble candle in a lantern.

Batten died in 1667, willing the Harwich lighthouses to his heirs, and a legacy to Mingo, of twenty pounds a year for the rest of his life for his keeper duties at the Harwich lighthouses. Other than knowing he was a servant of Batten, possibly enslaved, and dedicated to keeping the light, therefore saving many sailors, Mingo's life remains a mystery to us.

Willis Augustus Hodges - Cape Henry Lighthouse, Chesapeake Bay, Virginia

In 1870, Willis Augustus Hodges, the first African American lighthouse keeper at Cape Henry, didn't just tend the light—he carried a legacy of freedom, education, and activism as an abolitionist. He stands out in Maritime history as the first African American lighthouse keeper at Cape Henry, which is the 4th oldest lighthouse in the United States.

He lived in the keeper's quarters with the three assistant keepers he supervised.

William Roscoe Davis -Old Point Comfort Lighthouse, Hampton Roads Harbor, Virginia

William Davis and his family were among the first enslaved African Americans who sought refuge by the U.S. Army when they occupied Point Comfort in 1861. During the Civil War, most east coast lighthouses were damaged, destroyed or put out of commission. But not Point Comfort. The lighthouse remained undisturbed and functioning because Fort Monroe stayed under Union control.

By 1870 Davis the official head keeper of the Old Point Comfort lighthouse and held the post for about a decade.

When he was keeper, African Americans weren't allowed on public beaches, but Davis kept the sandy shore in front of the lighthouse open to anyone who wanted to use it. He ensured the light guided mariners into Hampton Roads Harbor, where the James, Nansemond, and Elizabeth rivers converged. William Davis' service illuminated the waters and the path toward equality.

William Parker Killock Shoal Lighthouse, Chincoteague Channel, Virginia

In February 1886, William Major Parker, an assistant keeper at Assateague Lighthouse for many years, was promoted to principal keeper at the newly built Killock Shoal Lighthouse. For 19 years his life there was routine, handling all his lighthouse duties with his wife, Venus, assisting him until his standard supply run to Chincoteague Island in October 1905.

His story took a dramatic turn. There had been a shooting, a murder, earlier that day on the island. The local law was rounding up a pose and demanded Parker join it. Bound by federal regulations that did not permit him to leave the lighthouse for an extended period, he had to decline. So, he thought nothing more about it.

On his next routine supply run to Chincoteague Island he was restrained, jailed for not joining the posse. The Lighthouse Board defended Parker.

Here is the 1905 article from The Alexandria Gazette: *"It is the case of William M. Parker, keeper of the Killick lighthouse on Chincoteague Bay, Va., and the question to be determined is to what government he owes first allegiance, the federal or state. Some weeks ago, Parker was summoned to join a posse to seek out an alleged murderer. To go on this task meant that he forsakes the light. He concluded it was better to save lives than revenge them, so he stole off to the lighthouse. A month later when he came ashore for his mail and provisions, he was arrested for failing to join the posse. His trial will soon be heard, and the Department of Commerce and Labor has requested the Department of Justice to defend him. This will be done."*

The outcome was full vindication. Instead of slumped in the cell, feeling like an iron anchor was holding him down, he was sighing with relief and holding his head tall. Local officials dropped the charges. William Parker was able to return to his work and his wife awaiting him at the Killock Shoal Lighthouse.

Parker kept the light shining at Killock Shoal for 25 years, until January 23, 1912, the day the lighthouse went dark. His wife Venus was on shore visiting friends. She returned to find her husband on his knees, his head laying on the bed—dead.. William Major Parker died at age of 57. He'd dedicated 36 of those years to lighthouse service.

His widow, Venus, took over his duties until a new keeper was appointed. Her story is listed above under women lighthouse keepers.

The maritime expertise of indigenous people, especially those from coastal tribes, was rooted in centuries of tradition. And many Native Americans served at lighthouses, though seldom recognized as official keepers.

Old Point Comfort Lighthouse, Cape Comfort Peninsula, Virginia

Before William Roscoe Davis was a lighthouse keeper there, before any lighthouse was built on that spot, way back in the 1500s, there are records of Native Americans keeping wood fires burning at the point to guide Spanish ships.

The Makah Nation- Cape Flattery Lighthouse, Strait of Juan de Fuca in Washington state,

The Cape Flattery Lighthouse was built in 1854 and stands within the Makah Indian Reservation on Tatoosh Island. It's named after for Tatoochatticus, usually referred to as Chief Tatooche. We have this 1788, historical line written by English fur trader John Meares describing the chief: "Canoes came out from the island with twenty or thirty men in each who looked very savage, with painted faces and sea otter cloaks. They were armed with bows and arrows with barbed bone points and large spears pointed with mussel shell. The group included the chief, Tatooche, whose face was painted black and glittered with sand".

The Makah were highly skilled mariners, using sophisticated navigational and maritime skills. They traveled the rough waters of the Pacific Ocean and the swift waters of the Strait of Juan de Fuca with relative ease. For centuries the Makah relied on the bounty of the sea for survival, from seals to salmon to whales. The sea is still a large part of the Makah's livelihood and way of life.

Members of the Makah Nation are therefore ideal at ensuring safe passage for ships navigating the milky fog, perilous currents, and surging tides at the strait. Here is a historical line from 1778 written by the infamous Captain James Cook, describing the strait of Juan de Fuca, "Beetling cliffs, ragged reefs, and huge masses of rock cut by the waves abound on every side."

The Ojibwe Nation-Wisconsin Point Lighthouse

The Ojibwe people, also known as the Chippewa, lived on the lands around the lighthouse, fishing and hunting for centuries. The "Witches of November" is what people in the Great Lakes area call the shrieking winds lashing Lake Superior at a speed of more than100 miles an hour. The Ojibwe's vast knowledge navigating the winds and other weather patterns, dangerous reefs, and seasonal changes was vital to lighthouse keepers.

Their legacy of living in harmony with the land and ecosystem, honoring it as sacred and essential to their own well-being, was crucial to the success of a recent restoration project. It involved removing invasive species and replanting native vegetation to return the land to its original beauty and vitality.

But at one time, the U.S. tried to force the Ojibwe from this land in Northern Wisconsin and the Upper Peninsula. In 1852, Kechewaishke (Chief Buffalo), principal chief of the Lake Superior Chippewa, journeyed to Washington, with a birchbark petition adorned with a crane, a bear, a marten, a catfish, and a man. Each line from heart to heart, eye to eye, symbolized the unity of the Ojibwe tribes. Standing before President Millard Fillmore, Chief Kechewaishke presented the petition and his resolve spared the Ojibwe from removal, securing their right to a permanent reservation in Wisconsin.

The Ojibwe Nation- Point Iroquois Lighthouse, Chipwa County, Michigan

Brave Ojibwe won a great battle against the Iroquois at the point in 1662—that is where it got its name. In the Ojibwe language—Anishinaabemowin, a branch of Algonquian, the name for the Point, *Nau-do-we-e-gun-ing*, means "Place of Iroquois Bones."

The lighthouse stands on a bluff above Lake Superior, which in Anishinaabemowin is called *gichi-gami* and means great sea. In the poem, The Song of Hiawatha, Henry Wadsworth Longfellow spelled it as “Gitche Gumee".

Though Hiawatha was a fictional character, he is based in part on Ojibwe oral traditions of a great warrior named Nanabozho. As “The Song of Hiawatha” implies, the Ojibwe people have a long history of navigating and fishing along Lake Superior in their birchbark canoes and they know these waters better than anyone.

Ojibwe Nation- Split Rock Lighthouse, Minnesota

The Split Rock Lighthouse stands sentinel on Minnesota’s rugged coast. The lighthouse itself seems like an extension of the rock and water that the Ojibwe have an enduring ancestral connection with, reflected in their work of assisting lighthouse keepers or as lighthouse keepers.

Mishi Peshu is a type of Loch Ness Monster but much more, which is said to lurk beneath the surface of Lake Superior. The Ojibwes’ belief in this lynx-like water-beast, its name means Great Lynx, demonstrates their strong spiritual connection to the Great Lakes, and the respect they hold for the lakes’ power and mystery.

It lurked beneath the surface of Lake Superior. Before any

warrior set out in their birch-bark canoe they paid homage to the lake's Great Spirit.

Many Hispanic families worked in lighthouses in Florida, Texas, California, and Puerto Rico.

Juan Andreu–St. Augustine Lighthouse, Florida

St. Augustine as the oldest port in the U.S., has many achievements, including the first lighthouse on the eastern coastline. In 1824, only three years after the U. S. purchased the Florida territory, Juan Andreu became St. Augustine's first official lighthouse keeper. He was born in Florida on the New Smyrna Settlement, a plantation, in 1774, to parents who voyaged here as teenagers from a Balearic Island off the Spanish coast. They were known as Menorcans and most likely came here with a large group to work as indentured servants on that plantation.

As head lighthouse keeper, Juan Andreu was the first Hispanic American to serve in the Coast Guard, and to manage a federal facility.

For two decades, with a break in between, 1824-1845 and 1854-1859, Juan Andreu and his family kept the light burning on Anastasia Island—an isolated strip of land.

His wife, Maria Mestre de los Dolores Andreu, picked up the mantle of assistant lighthouse keeper. In 1845, one of Juan Andreu's cousins temporarily took the job until he and his family returned to their lighthouse duties in 1854.

Tragedy struck on December 1859, when Andreu stood on a homemade scaffold, painting the tower. Abruptly, the platform gave way. Juan Andreu plummeted 150 feet to his death.

His wife, Maria Andreu, was then appointed the official lighthouse keeper, see her story in the women lightkeepers section.

Robustiano Rivera - Guánica Light —Faro de Guánica, Puerto Rico

Across the vast ocean, in Faro de Guánica in Puerto Rico, in 1898, during the Spanish American War, when the island was under Spanish rule, Robustiano Rivera, principal lighthouse keeper at Guánica Light, gasped as he spied an American military convoy approaching the Guánica Bay.

Wasting no time, he ran through Guánica, alerting everyone that the Americans were coming. And he kept going, marching to Yauco, he reported the invading forces to the mayor.

Probably due to Robustiano Rivera's warning, only seven lives were lost when the Americans marched on Puerto Rico. The U.S. flag was raised by mid-August, the Spanish-American War ended with the signing of the Treaty of Paris in December that year, and Puerto Rico became U. S. territory.

PART II: REMARKABLE LIGHTHOUSE KEEPERS: WOMEN

JULIET NICHOLS - POINT LIGHTHOUSE, on Angel Island, San Francisco Bay, California

In 1902, a widow, Juliet Nichols, took the position of the keeper of Angel Island lighthouse in San Francisco Bay, fully understanding the meaning of service and sacrifice.

Four years earlier, her husband, Captain Henry Nichols, gave his life for his country in the Spanish American war.

The lighthouse perched on a small, rocky point on Angel Island was equipped with a Frensel lens, casting a bright, steady beam over the treacherous waters of San Francisco Bay. The wind-up clockwork mechanism paired with the lens, signaled a hammer to strike a massive bronze bell.

The echoing clang guided mariners through the cocoon of fog that wrapped the bay.

July 2, 1906, the fog was so thick it was as if a cloud swallowed the shoreline when the bell's wind-up bell mechanism shuddered to a halt. The silence portended disaster for the mariners. How would they make it to shore?

Grabbing the disabled hammer, Juliet swung in the clockwork's rhythm, striking twice every 15 seconds. Cold metal vibrating under her fingers, clenched tight around the hammer's handle, the weight of it bearing down on her arms, rattling her bones with every manual blow. Hour after hour. Through the night. Relentlessly. After 20 hours and 35 minutes of Juliet's intense labor, the fog lifted.

The next day, a repairman tinkered with the mechanism, assuring Juliet Nichols he'd fixed it.

But the next day, July 4th, it failed again. This time, the hammer broke. It was a life-or-death situation for mariners entering the bay. Juliet had to find a way to ring that fog bell. With a nail hammer in hand, she swung that small metal head, ringing that fog bell twice every 15 seconds. Sweating from exertion, hands raw from rubbing against the handle upon each jarring impact with the bell, and her body shaking from exhaustion, she pressed on. Because each clang served as a lifeline to the ships lost in the thick fog, leading them to the safety of the harbor. She didn't stop until the break of dawn when the fog vanished.

Juliet Nichols achieved a commendation from the Lighthouse Board—for her bravery and unyielding commitment to the mariners she served. She proved ingenuity, sacrifice, and selfless commitment resonate louder than any fog bell or clockwork mechanism.

Kathleen Moore - Black Rock Harbor Lighthouse, Bridgeport Connecticut

Kathleen Moore, often called Kate, maintained the lighthouse's beacon at Black Rock Harbor Light.

Her father became lighthouse keeper there in 1817. At age 12 she began assisting him, walking on two slippery planks submerged four feet under water during raging storms, her whale oil lantern swaying in her small, wet hands. As her father got older and could no longer care for the lighthouse, she handled all duties, including keeping the light burning, and caregiving for both her parents.

In 1871, at the age 59, the U.S. government officially appointed Moore lighthouse keeper. A task she had already been performing for 47 years. She is one of the first female head light keepers in U.S. history.

Kathleen Moore had to stay up with no sleep on stormy nights because the howling wind lashing Black Rock Harbor would put the light out. So, she slept alone in the lighthouse, beside the flame to keep it burning and guiding mariners safely to shore. Her vigilance was crucial.

A historical quote from Moore is, "Sometimes there were more than 200 sailing vessels at night, and some nights there were as many as three or four wrecks." She is credited with saving the lives of twenty-one mariners. And she nursed many of them back to health.

As diligent as she was in guarding the light and rescuing sailors, Moore was just as fierce in guarding the oyster beds from poachers, rowing out to them with her loaded shotgun, warning. "I represent the United States Government, and you have to go."

When Moore was 66, a reporter with the New York Sun interviewed her, writing that her walls were lined with book-filled shelves and wooden duck decoys she'd carved herself and sold to visitors. Also, she'd inherited an original Peter Paul Reubens painting from her mother's family that hung on the wall of the small, humble Keeper House.

In 2010, the U.S. Coast Guard named the ninth Sentinel-class cutter, Kathleen Moore, in her honor. With its engines humming, it glides by the tropical cost of Key West, Florida, carrying her legacy forward.

Grace Darling: The Longstone Lighthouse on Brownsman in the Farne Islands, Northumberland Coast, England.

Grace was the daughter of William Darling, head keeper of Longstone Lighthouse on Brownsman, one of the Farne Islands.

On September 7, 1838. during a violent storm, she spotted the steamer, the SS Forfair, shipwrecked a mile from the lighthouse. Swirling, howling winds and towering waves had crashed the ship onto a rocky shoal by Big Harcar Rock. At Grace's urging, her father relented to risk a rescue in the hammering rain, surging tide, whipping winds, and churning sea during the rough storm. They shoved off in a twenty-foot wooden rowboat that teetered on the brink of capsizing with every swell. Grace rowed with all her might as winds and waves rocked the small-framed boat.

Rising waves pounded the frail rowboat relentlessly as Grace and her father struggled to haul five of the nine survivors aboard.

They managed to make it back to the lighthouse, where Grace tended to three of the men she helped rescue. Her father and the other two survivors braved the tempest to return to the rock and save the other four. Hailed as a national hero, Queen Victoria wrote Grace a letter of admiration, the Humane Society awarded her a gold medal, and she received £750. At the young age of 26, in 1842, tuberculosis took Grace. Likely the prolonged exposure to harsh, frigid rain, and tempest winds, affected her lungs. But Grace Darling used the years she had well, saving lives and inspiring others.

Minnie Paterson: Cape Beale Lighthouse, Vancouver Island, Canada

December 6, 1906, Minnie Paterson, the wife of Cape Beale Lighthouse keeper, Tom Paterson, woke to 80 knot winds roaring through the rugged cliffs and rain battering the small keeper's house.

In the distance, she and her husband spotted a schooner, the Coloma, run aground on a reef—masts splintered, and the crew clinging to the wreckage. The storm had severed the telegraph lines.

The lighthouse tender ship, the Quadra, six miles away, was

the survivors' best chance of rescue. Minnie took off on a six-mile trek, wading thorough flood water and storm debris across a sand spit connecting the cape to the mainland. In the pelting rain, she grasped the wet, telegraph wire strung between trees, to keep her footing. Four grueling hours later, she reached Bamfield Creek, but the rowboat she was going to take to the Quadra was missing.

With lives to save, she kept going, heading to the home of her friend Annie McKay, the wife of a former keeper. The two women dragged a skiff to the water. The turbulent winds almost turned the skiff over, but they kept rowing through the churning inlet to the Quadra where they reported to the captain. He set sail to the schooner on a rescue mission, Minnie and Annie, clothes sodden, and legs cramping, didn't stop. They climbed the hill to the telegraph station to get more help.

Minnie could barely stay on her feet as she slogged her way through the storm. Annie had to catch her at one point to keep her from falling as hypothermia set in—her mind foggy, her body racked with chills. Minnie made it back to the keeper's house to her husband and children. Exhausted to the bone, she collapsed into a chair. However, she made sure it was the chair in front of the telescope. Minnie peered through it as she had to watch as the Quadra reached the wreck. The tender ship made in just in time. Moments after they saved the crew, the Coloma broke apart on the reef.

Minnie was awarded a gold watch and a silver plate by the Canadian Government for her heroism, and the Seattle Maritime Union presented her with a silver tea set. Yet per prolonged exposure to the elements during her grueling journey to save the crew of the Coloma weakened her lungs, and five years later, Minnie died of tuberculosis.

Minnie Paterson's courage, sacrifice, and dedication shines through history like the beam of the Cape Beale lighthouse.

Ida Lewis: Lime Rock Lighthouse, Rhode Island, USA

Captain Hosea Lewis's daughter, Idawalley Zoratia Lewis,

went by Ida Lewis. Her father was appointed keeper of Lime Rock Lighthouse in Newport, Rhode Island, ensuring the beacon burned brightly to guide ships through Newport Harbor. But the Lime Rock lighthouse didn't have keeper quarters, the Lewis' had to live in town, and the captain commuted to work by boat each day. Ida usually went with him, helping with the lighthouse duties.

But his work came to a stop when a stoke left him paralyzed. The lighthouse board stepped in and built a keeper's house on Lime Rock. Once the Lewis' moved in, Ida, only 15, and her mother, took over the lighthouse duties.

Ida's rescues are legendary—because she risked her life so many times to save others. Her first rescue was on September 1858. A prank went awry for four schoolboys. Ida heard their laughter turn to screams. The small boat capsized, tossing them into the changing tides and depths of New Port Harbor. Their heavy clothing dragging them under. The children's flailing arms and kicking legs splashing wildly as they fought to keep their heads above the waves.

Ida rushed to her small boat, rowed to them, and pulled one boy after the other aboard. Her muscles strained from hauling the boys in and rowing back against the tide. At the lighthouse, her mother tended to the rescued children with warm blankets and steaming cups of tea.

Then there're the three men and the runaway sheep. A local banker's prized sheep bolted into the harbor. The men rowed after it, but their skiff capsized. Hearing their cries for help, Ida leapt to action. In the daring rescue, she saved the three men and took them to shore. Then rowed back in pursuit of the sheep. She lassoed it and towed it back to land.

By 1869, Ida had saved the lives of nine people and one sheep. Her most famous rescue took place in March of that year. A storm broke across the harbor, taking the life of a 14-year-old boy hired by two soldiers to ferry them from Fort Adams into the harbor. He drowned when the storm flipped the boat over. The soldiers managed to cling to the hull of the capsized boat. Ida dashed out shoeless with her younger brother, Osa. They powered through the white cap waves slamming into the skiff and the pelting rain. Osa helped her haul the stiff, shivering, freezing men aboard the skiff. After all, Ida only weighed a little over 100 lbs. Her family cared for the men until they recovered.

A reporter from the New York Times wrote an interview on her haling Ida Lewis as "The Guardian Angel of Newport Harbor." Newport declared July 4th, 1869, Ida Lewis day and through local donations the city presented her with a new rowboat—named Rescue. Some of the celebrities who visited her included the suffragettes Elizabeth Cady Stanton and Susan B Anthony.

As for what Ida thought of this recognition, she told a reporter, "If there were some people out there who needed help, I would get into my boat and go to them even if I knew I couldn't get back. Wouldn't you?"

After years of unofficial service, Ida became the head keeper of Lime Rock Lighthouse in 1879. Another of her more famous rescues happened on February 1881.

Two soldiers were walking across the frozen harbor from Newport to Fort Adams. Ida watched from her window as the ice gave way, plunging them into the frigid water. She grabbed a clothesline and raced onto the ice with her brother Rudolph, the assistant keeper, following. Standing on firmer ice, Ida tossed them the clothesline. She and Rudolph tugged the men to safety and took them back to the lighthouse to recover until they got them to the fort's hospital.

For that act of heroism, the U.S. government awarded Ida a gold lifesaving medal—the first woman to receive such an honor. Another of her medals was from the Massachusetts Humane Society for her heroic and successful efforts in saving human lives.

In 1889, Ida's seventy-year-old uncle Henry Lewis was rowing back from a fishing trip and fell overboard near the lighthouse.

But his niece, Ida, saw the accident and rushed to him, saving her uncle from drowning.

Ida kept performing rescues even in her late sixties. On August 4, 1909, five girls were rowing around Newport Harbor when waves from The Commonwealth, a 425-foot steamship, headed from the harbor to New York, swamped the girls' boat, throwing them into the water. Sixty-eight-year-old Ida climbed into her boat, the Rescue, rowed to the girls and pulled them in.

Ida is credited with saving 18 lives. In 1911, at 69, stricken by a stroke, she died three days later. Her epitaph reads: "Ida Lewis, the Grace Darling of America, Keeper of Lime Rock Lighthouse." In 1924, the lighthouse was renamed the Ida Lewis Lighthouse, the only American lighthouse to bear the name of its keeper.

St. Simons Island Lighthouse Georgia

From 1836 we have a documented account of a head keeper who got gout and turned his duties over to an elderly, enslaved Black woman while he was sick. This woman, who served the mariners, was not paid for her hard work and sacrifices, she received no awards or recognition of any kind. We don't even know her name. But in steadfastly performing these duties put upon her; in keeping the light burning, she guided mariners around the treacherous sand bars of St. Simon Sound. We know they and their families were deeply grateful she was there. Otherwise, the light would surely have gone out and they may have not made it to shore.

New Point Comfort Lighthouse Virginia

This was a similar situation to the woman mentioned above. An enslaved, African American woman, her name lost to history, served as an assistant keeper to the Head Keeper, Isaac Foster, at the New Point Comfort Lighthouse in Virginia, around 1852. Her efforts saved sailors and kept the light burning.

There were many other enslaved Black women and men who performed the duties of the official keepers who lived far away or were ill. Though we may never know their names, they all served a crucial role in U.S. maritime history, keeping the light shining and saving lives.

Venus Parker— lightkeeper at Killock Shoal Lighthouse

Venus Parker married a lighthouse keeper, William Major Parker, and moved into the square, four room, one and a half floor keeper cottage at Killock Shoal Lighthouse. As was typical of these lighthouse keeper wives, unofficial and unpaid, she took on the duties of a keeper's assistant.

Many years later, January 23, 1911, when she went to visit friends on the mainland, she returned to find he had died on duty, while praying on his knees at his bedside. Friends had to carry her husband's body to the mainland in her stead because she had to stay and keep the light burning as he would have wanted. Mariners' lives depended on her. She continued to tend to the lighthouse for a more than a year until another keeper was finally appointed in March 1912. To her husband's credit, Venus made sure the light never dimmed on her watch.

In 1824, Juan Andreu became St. Augustine's first official lighthouse keeper. His wife Maria Andreu was assistant lightkeeper for her husband. But in December 1859, when Andreu was painting the tower the scaffold he rigged broke. He died instantly from the 100-foot fall.

His wife, Maria Andreu, was then appointed the official lighthouse keeper, achieving the status of the first woman to serve in that position in Florida and the first Hispanic American woman to manage a federal shore facility. She proudly kept the light burning until 1862, when the Civil War brought her duties to a halt. At that time, the light was extinguished so it couldn't be used to aid the Union Navy.

So, with that, let's move to the next chapter — Lighthouses During War.

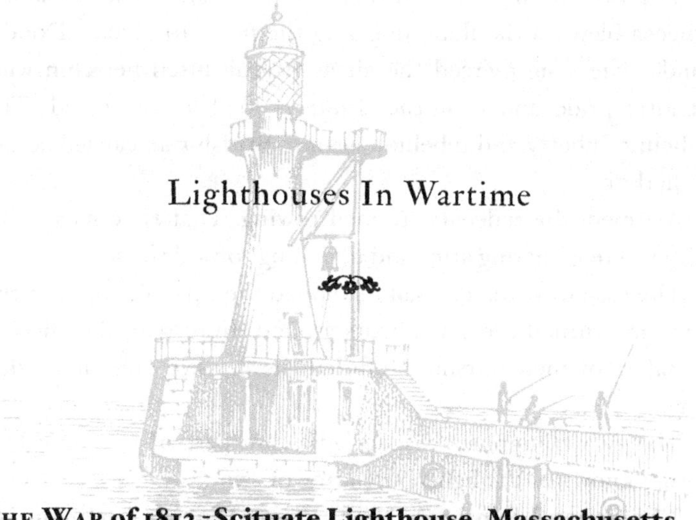

Lighthouses In Wartime

The War of 1812 - Scituate Lighthouse, Massachusetts

June 1814, two British ships sliced through the water of Scituate Harbor. Raiding parties spilled onto the docks, setting American fishing boats aflame and stealing others. Simeon Bates, lighthouse keeper and soldier in the militia, fired the brass cannon, the booming shots chased the British out to sea.

In September, the threat returned. A sleek British warship, the *La Hogue*, loomed offshore. Bates wasn't at the lookout point. It was unmanned at that moment. No soldiers were there to mount a defense or alert the town and save everyone. Scituate Lighthouse was empty except for three women: Bates wife, and two of his daughters, 21 old Rebecca, and 17 old Abigail. They had a musket. But they wouldn't be able to kill more than two men before they were shot dead are taken by the British. What could they do?

Use their wits. Trick the British. So, instead of a gun, Rebecca grabbed the Fife and drum the militia kept at the lighthouse. She and Abigail dashed behind a ridge of cedars. Concealed by the thick foliage, Rebecca raised the fife to her lips and her sister poised her drumsticks.

The British sailors rowed their landing crafts closer to shore. Rebecca blew on the flute, fingering the notes to Yankee Doodle Dandy. The song pierced the air as Abigail lifted her chin with patriotic pride and beat the drum as loud as she could. The anthem of liberty and rebellion against British rule carried across the harbor.

Alarmed, the redcoats stopped rowing. That tune meant the militia were gathering arms and marching toward them.

The ruse worked. The sailors dipped their paddles back in the water and turned the small boats around. In retreat, they fled to the safety of their warship, unfurled the massive sails, and sailed away.

The two triumphant young ladies are part of maritime lore and are known as the lighthouse army of two.

The Second Seminole War—Cape Florida Lighthouse, Key Biscayne in Miami-Dade County, Florida

The lighthouse keeper, John Dubose, headed to Key West for supplies, leaving his assistant, John W. B. Thompson, in charge of the lighthouse that guided mariners through the dangerous Florida Reef and violent gales of Key Biscayne. Around 4:00 pm, July 23rd, 1836, as Thompson walked to the keeper house, a war band of fifty Seminoles, no more than twenty yards away, gave chase.

Luckily, he wasn't alone. As Thompson sprinted to the lighthouse, he yelled, "Run for your life," to Aaron Carter, his elderly Black servant, possibly enslaved.

The Seminoles fired a volley of musket balls that tore the men's clothes and riddled the door with buckshot. They made it inside just in time, but as Thompson turned the key, a warrior grabbed the door.

But Thompson managed to lock it. He grabbed three loaded muskets he kept on hand. Carter guarded the door as Thompson dashed upstairs. The war painted Seminoles, yelped bloodcurdling war cries as they surrounded the keeper's house.

Thompson fired his musket, three shots a minute out the second window, reloaded fast as he could, darted to another window, and blasted more musket balls at the enemy. The warriors smashed the windows and returned fire. The loud crack of each shot and the sound of shattering glass had Thompson's ears ringing. The gun fight went on until night fell.

The Seminoles heightened their attack, letting loose a fierce volley, rushing the lighthouse door, and setting the boarded windows ablaze. The yellow pine wood burned fast. Thompson grabbed a keg of gunpowder, a musket, and lead balls and dashed up the stairs to the top of the house with Carter at his heels. Seminole musket balls pierced the tin tanks, spewing 225 gallons of whale oil into the flames. The inferno exploded with crackling flames rising around them. Forcing the men to retreat to the two-foot-wide iron platform outside. They lay flat on their backs, lamps bursting, shards flying in all directions, their clothes on fire, skin burning, warriors firing up at them. Carter called out that he was wounded. Those were the last words he ever spoke.

Aaron Carter, seven musket balls lodged in his body, died on duty, repelling the Seminole attack. The first U.S. Lighthouse employee killed in combat, an African American who served and gave the ultimate sacrifice to his country in 1836.

Desperate to end his torment, Thompson tossed the keg of gunpowder into the roaring flames to blast everything, including himself, to dust. The explosion rocked the tower, top to bottom, but didn't blow him into eternity as he'd hoped. However, the landings to collapsed, tamping down the fire, giving him a moment's relief. A second later, it blazed as hot and wild as ever.

Musket fire like hail in a storm fell around Thompson. Wounded with three musket balls in each foot, he saw no way to survive. He was prepared to launch himself beyond the iron railing, into the air, but didn't have the strength.

That night, the Seminoles, sure he was dead, plundered then torched the keeper's house. At about 10 the next morning a group of warriors loaded Thompson's sloop with booty and left. The rest headed to the other end of the island.

That afternoon, Thompson spied two ships and drew on all his strength to signal them by waving Carter's bloody pants. Captain Armstrong heard the explosions and brought a detachment of seamen and marines to investigate. On the way, they recovered Thompson's slope which had been stripped by the Seminoles.

The marines rescued Thompson by tying twine to a ramrod and shooting it from a musket to the top of the tower. He managed to grab the twine, fasten it to an iron stanchion, pulled up a 2-inch-thick rope, and tied it to the stanchion. Two marines shimmied up the rope and Jerry-rigged a sling to haul Thompson down.

Thompson recovered after a short stay in Key West, where they were able to remove all but one of the musket balls in his feet. Seminole war parties continued to raid the area, which caused delays in repairing the lighthouse. It wasn't operational again until 1846.

By 1858, the Seminole nation moved deep into the Everglades to evade Army officials who wanted to remove them from their homeland of Florida. Many still live there.

American Civil War _ Point Lookout Lighthouse, Maryland

Pamelia Edwards began as lighthouse keeper in 1855 and continued her service when the Civil War broke out in 1861.

Pamelia Edwards kept the light burning during the turbulent war and several years after until 1869.

The Point lighthouse stayed in the hands of the U. S. government who also leased the land on Point Lookout to build Hammond General Hospital for sick and wounded U. S. soldiers.

It was one of the largest and busiest Union war hospitals. Unfortunately, many of the men did not recover and died there.

On a brighter note, the hospital staff helped several Black men and women, enslaved in Maryland, escape to freedom.

The federal government also built a POW camp on Point Lookout, imprisoning a total of 50,000 men between 1863 and 1865.

The camp conditions were deplorable. They housed the confederates in nothing more than overcrowded tents on a flat stretch of sand inside a walled pen by the shore, subjecting them to scorching heat and turbulent coastal storms.

Nearly 4,000 of the prisoners died from disease, exposure, and starvation. African American U.S.C.T. Regiments served as guards at the camp. The Veteran Reserve Corps also served as camp guards, as well as orderlies and stewards in the Union hospital. Camp officials recruited some of the prisoners into the First Regiment and Fourth Regiment U.S. Volunteers and sent them west to battle Native Americans.

In 1864, Confederate General Bradley T. Johnson planned a raid to liberate the prisoners, arm them, and march on Washington. But Union authorities got wind of it and the mission was abandoned.

WWI — Fowey Rocks Light Station, near Cape Florida on Key Biscayne.

During the war, three or more military signalmen were assigned to each chosen lighthouse. There were some instances of friction between the Navy Signal men and Lighthouse personal. Let me tell you some things that happened at the Fowey Rocks Light Station, which replaced the Cape Florida Lighthouse of the Seminole Wars.

As the Lighthouse inspector's boat drew near the tower, a Navy Signalman stepped out on the balcony of the lighthouse and urinated down towards the inspector. He told the signalman that the lighthouse had a toilet, and he should use it.

The signalman, out of uniform, dressed only in dungarees,

asked if his urination would hurt the ocean. What's humorous about the incident is the spelling error on the inspector's report—*naked to the waste.* A telling error, considering what the signalman had been doing.

On November 11, 1918, at the eleventh hour, a cease fire was called. The Lighthouse Service returned to its prewar home, the Department of Commerce, where it remained until 1939.

WWII — St. Catherine's Lighthouse and The South Fair Isle Lighthouse, England

During the war, some lighthouses were completely extinguished while others were merely dimmed to aid merchants and fighting ships. Fighter planes and sea battles could be observed from the lighthouses. The military also used them as signal station, weather station, and intercepting station. Both sides painted some lighthouses with camouflage for safety. Also, both sides surrounded some of their lighthouses with barbed wire fencing and land mines.

There were some lighthouse casualties. In the UK, St. Catherine's Lighthouse, mentioned in chapter 1, built in 1323, on the southernmost tip of England, on the Isle of Wight, was bombed in 1943, killing three lighthouse keepers.

Also in the UK The South Fair Isle Lighthouse in southwest England was bombed twice. First, in December 1941, killing the assistant lightkeeper's wife. Six weeks later, in January 1942, the lighthouse was hit again, taking the life of the principal keeper's wife and daughter.

Many lighthouses have not only stood against the tumultuous seas but also against human foes during major conflicts and world wars. Point Lookout Lighthouse in Maryland has stood up to the shifting currents, rising and changing tides, and the churning water and treacherous wind gusts of sudden storms as well as the Civil War. But that's not all, it has also weathered a great deal of paranormal activity. Many visitors and employees experienced hauntings there. A team of paranormal investigators researched the lighthouse and identified numerous findings of unexplained phenomena.

Next, we'll explore lighthouses invaded by drowned seamen and other specters.

Haunted Lighthouses

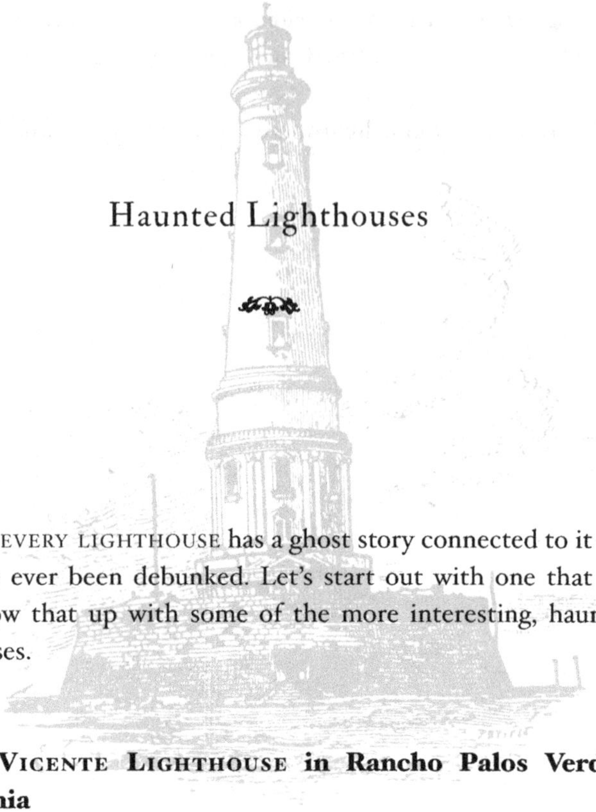

ALMOST EVERY LIGHTHOUSE has a ghost story connected to it and few have ever been debunked. Let's start out with one that has and follow that up with some of the more interesting, haunted lighthouses.

POINT VICENTE LIGHTHOUSE in Rancho Palos Verdes, California

Almost nightly, keepers or visitors saw the ghost of a woman garbed in a flowing white gown. The story was she'd walked the grounds of the light station for decades, waiting for her sweetheart, who'd drowned in a shipwreck, to join her.

A young assistant keeper noticed she only appeared at night when the station's rotating beacon moved in her direction. Her appearances also occurred during light fog. Therefore, he concluded slight imperfections in the third-order Fresnel Lens caused the feminine ghost. As the lens rotated, it refracted light toward the ground in a confusion of arcs. If the refractions came together in just the right way and found a patch of fog, the lady appeared. But most visitors prefer to believe the apparition is a woman from the grave haunting the station grounds, searching for her sweetheart. And maybe it is.

The most unique thing about the Point Vicente haunting is the questionable activity was reasonably explained. Most lighthouses are said to have ghost and hardly any have been debunked.

Cabo Blanco Lighthouse, Puerto Deseado, Santa Cruz,

As hauntings go, this one is mild, but it's still one of my favorites. You see Cabo Blanco has a typewriting ghost. As a writer, I just love that.

In the 1950s, the lighthouse keeper was a Navy ensign, who preferred typing to handwriting because it was faster, neater, and easier to read. One day he was found slumped over the typewriter in terrible pain. The other keeper raced for help, but by the time he got back, the ensign was dead.

The old lighthouse is empty now and only shares the coast with seals and huge cormorant birds

But before it was closed, adventurous souls would brave this desolate spot and enter the red brick lighthouse, shocked to find the silence of the empty structure interrupted by the clicking of typewriter keys. Though nothing was there. Not even a typewriter.

Yaquina Bay Lighthouse, Lincoln County, Oregon

Muriel Travennard was a local teen who lived with her father. One day he had to leave on a two-week trip. But months later, he hadn't returned. She'd hadn't heard from him at all. Hoping to cheer her up, some other teens invited her to join them in exploring an abandoned lighthouse.

Muriel's father had been a sea captain, and often took her sailing with him on his sloop. So, the lighthouse intrigued her, and she went along.

The teens soon grew bored as the lighthouse was nothing but shambles. The only interesting thing they found was a strange iron door on the second floor that opened to a deep hole. When the group was ready to leave, Muriel realized she'd dropped her handkerchief inside and ran back for it. Her friends waited until it got dark. Worried, they called out to Muriel. She didn't answer.

Frantically, they searched the lighthouse. They didn't find Muriel but noticed a pool of fresh blood at the foot of the stairs. It hadn't been there before. Crimson droplets led to the next floor and all the way to the edge of the iron door. This time, the teens couldn't get it open. Terrified, they ran for help.

Officials conducted a thorough search, but no one could pry the compartment open. Even a crowbar couldn't budge it. The iron door remained sealed, as if guarding the secret of what happened to Muriel.

A rumor spread that a drifter hiding in the compartment, preyed on Muriel when she accidentally came upon him. To this day, dark stains mark the spot where she bled out that tragic night. The mystery was never solved. Muriel's body has never been found. The truth remains hidden behind the impenetrable iron door.

But many people have seen Muriel's ghost. She haunts the lantern room, so the tower is kept cornered off. But sometimes her lonely spirit floats to the second floor. Perhaps Muriel is waiting...for justice...or her father, who never came. And no one ever found out what happened to him.

Mintos Ledge Lighthouse, Massachusetts

The first lighthouse structure, erected in 1850 on a rock outcropping, about a mile from shore, offered little resistance to wind and rain. It was a screw pump tower, anchored to the ledge with nine iron legs supporting the small house for the keepers. The iron legs needed more braces—some of the cross pieces had bent. It rocked and reeled with each wave that passed through the iron legs and each screeching gale that blew by it.

The head keeper, Captain John Bennett, informed the lighthouse service of the safety concerns for himself and his two assistants, John Wilson and Joseph Antoine. Still, the service wouldn't fortify it.

Wilson and Antoine often played a game in the tower, signaling to each other through the long stovepipe that ran from the lantern room down to the living quarters. It was useful as well, in emergencies, to summon each other.

In 1851, as Captain Bennett went to shore for supplies, one of the worst New England storms on record bore down on Minton's Ledge. With no boat, Wilson and Antoine were marooned in the lighthouse. As the lashing winds shook the lighthouse, the men frantically pounded the fog bell. Shortly after midnight, the light went out, and the bell fell silent.

At dawn, debris from the lighthouse began to wash ashore, including the bodies of the two assistant keepers.

The lighthouse service constructed the second Mintos Ledge lighthouse, tall and made of stone, in 1860.

From the beginning there were sightings of ghostly aspirations. Many sailors on passing ships saw Joseph Antoine's ghost, his clothes dripping wet, clinging to the exterior ladder on the lower part of the tower. He'd wave frantically, pleading for help and crying out warnings to stay away from Mentos Ledge.

The new lighthouse keepers also experienced paranormal phenomena inside the tower. At night, when one keeper was up in the washroom and the other in the kitchen area they heard tapping sounds. Investigations to discover the source of the tapping were met with failure. The noise seemed to come from everywhere and nowhere.

One day the new keepers we're eating at the kitchen table when cold air crept through the tower carrying damp drops of water and exhaling a soft whistle as it blew upstairs. Then the tapping began, even urgent and louder than before, filling the entire tower. They sprung from the table and raced upstairs one behind the other, stumbling, shouting, and cursing at the horrible commotion. They found the gallery door wide open. Swinging wildly on its hinges inviting the biting wind to invade the lighthouse.

The lighthouse keepers shut the door, secured the latch, and examined the lens and lamps. Fortunately, there was no damage. No one could explain how the door came open. That's when they realized the tapping was meant to alert them.

A member of the construction crew who built the new lighthouse shared that when working there, Captain John Bennett ranted about the toll of a fog bell, though none was ringing. And the Captain also complained that his dead assistants would not stop yelling at him to help them.

If spirits inhabit the tower, they are lonelier than ever, for there is no one to scare these days. Automated in 1973, the lighthouse now has a romantic reputation for flashing a 1-4-3 signal, meaning— I love you... I love you... I love you..., which delights couples on shore.

Seguin Island Lighthouse, near Georgetown Maine

This light station, established back when George Washington was president, has had over forty keepers through the years. Few of them could handle the isolation and harsh weather on the rugged island.

In the mid-1800s, a lighthouse keeper's wife had a hard time adjusting to the isolation of lighthouse living. The perpetual fog at Sequin kept her in a state of depression. What she missed most from the mainland was music: singing with friends, choir rehearsals, and the rich, joyful sound of the church's piano and organ.

To ease her boredom and depression, her husband had a piano delivered so she could entertain herself. But only one sheet of music made the trip across the water.

So, the keeper's wife played that one song repeatedly, often singing along. When her husband brought her new sheet music, she wouldn't have it. Bent over the piano, at all hours, the woman kept playing the same tune, over and over again.

Residents along the shore also heard the never-ending melody as the sea breeze carried the notes the across water.

The lighthouse keeper realized she'd lost her mind, and it was making him just as crazy. One day, he just couldn't take it anymore. Overcome by madness, he wrapped his fingers around his wife's neck and strangled her—putting an end to the song. Then he swung an ax at the piano, violently chopping it into kindling. Lastly, the man took his own life.

But the faint sound of phantom fingers gliding over piano keys, playing a haunting melody, drifts from the fog covered island of Seguin to the shore. A solitary tune, playing over and over.

Owls Head Lighthouse, Maine

Owls Head is haunted by several dedicated keepers so devoted to the safety of mariners, maintaining the lighthouse, and never leaving their post, they linger in spirit form, tending to the lighthouse after they die.

In the 1980s, when Coastguardsman Gerad Graham was stationed at Owls Head, his two two-year-old daughter Claire, awoke in the middle of the night to meet her father at the top of the stairs and said, "Fog's rolling in! Time to put the foghorn on!"

Those terms had never been used in front of her. Over the next two years, Claire described a man with a beard wearing a blue coat and seaman's cap. Was this a dedicated keeper from beyond the grave who appeared to Claire to ensure the foghorn sounded on time.

Other paranormal activity at Owls Head includes disembodied voices murmuring in hallways, mysterious orbs of light dancing in the black of night, and shipwreck victims, from centuries past, spotted at the edge of the cliff.

The next chapter starts with Owls Head again, but this time with happier tales—those of lighthouse pets.

Remarkable Lighthouse Pets

So far in the pages of this book, we've met human lightkeepers, lightkeepers during war, and ghost lightkeepers, but we haven't talked about lighthouse pets. Fur or feathered companions were so important to lighthouse keepers because it got lonely at these isolated locations. Some pets also pitched in, helping with rescues or other vital duties.

Spot the Spaniel—Owl's Head Lighthouse, Maine

From 1930 to 1945, the head lighthouse keeper, at Owls Head, Augustus (Gus) Hamor, had an assistant with four legs—a spirited Springer Spaniel, called Spot.

Hamor's daughters, Pauline and Millie, taught Spot the trick of pulling the fog bell rope with his mouth.

Anytime a ship passed, Spot bounded into action, tugging the rope. The clangs echoed over the white foamed waves to the ships, which replied with a toot or a toll from their horn or bell.

Spot loved the mailboat best and always greeted the skipper, Stuart Ames, with a wag of his tail. Because Ames always had a treat for the loveable dog. And with his keen canine hearing, Spot came to recognize the hum of the mailboat's engine before it arrived.

A fierce snowstorm rolled in, blanketing the peninsula in white. Caught in the blizzard, Stuart Ames desperately tried to steer through the swirling snow. Spot's keen ears picked up the whirr of the mailboat's engine. The dog scratched at the door until Gus let him out. Only Spot's head could be seen above the snow drifts as he struggled to the cliff below the lighthouse. There, he tugged on the icy snow-covered rope with his teeth. But with the fog bell covered in snow, it only made a muffled clunk.

Spot made his way to the bay's edge. His loud, incessant barking pierced through the harsh howl of the arctic winds. The faint sound of the mailboat whistle was heard in reply.

Spot's barking kept Ames from crashing his ship onto the rocky shore and guided him safely into the harbor. With his quick thinking and loud barking, Spot saved Stuart Ames' life.

The faithful dog lived out his days by Gus's side and was buried beneath the fog bell he loved so much. A true lighthouse hero.

Milo the dog—Egg Rock Lighthouse, Massachusetts

The first lighthouse keeper there, George Taylor, brought his Newfoundland/Saint Bernard mixed dog, named Milo with him there in 1855.

The family had an open space in the masonry of the entryway to the house so Milo could come and go as he pleased. In foggy weather, Milo would constantly bark at any ships approaching the lighthouse, becoming an additional fog signal. Milo assisted in the rescue of several children from drowning near the island. Perhaps even George Taylor's son, Fred.

Sadly, Fred Taylor really did drown in later years, Milo was not there at the time to save him.

Smut the dog—Two Bush island lighthouse, Spruce Head, Maine

Aldiverd A. Norton, the first keeper at Two Bush Island, from 1897 – 1909, owned a mutt named Smut. In 1900 a winter gale drove the fishing schooner Clara Bell onto the rocks near the lighthouse.

Nothing could be heard over the clamorous wind. But Smut's canine hearing picked up the grinding of the Clara belle's hull against the stone and the men's cries of desperation. He whimpered to be let outside and soon saw for himself that help was needed. He ran back to the house, wet and shivering, and scratched the door to be let inside. He began barking urgently scratching the door to be let out again.

The lighthouse keeper Norton was perplexed by the dog's behavior and realized something was amiss. He followed Smut outside and over the slippery rocks for some distance until he spotted a small, shallow boat with two men on board. The lighthouse keeper quickly got a rope and just as he was going to throw it to the men, the boat capsized. Norton struggled with the rope until the men caught it.

After he pulled them safely to shore, Norton told them how the dog saved their lives. The fishermen fell on their knees and hugged Smut. They tried to persuade the lighthouse keeper to sell Smut to them but Norton said, "No. Some dogs are priceless."

Sambo Tonkus the Cat — Nubble Lighthouse, an island off Cape Neddick, Maine

The Nubble lighthouse keeper Eugene Coleman and his wife had a ginger-striped tabby named Sambo Tonkus, that grew to a hefty twenty pounds.

Over the years, Sambo Tonkus made the Nubble his home. He stayed there as head keepers came and went. The cat was a permanent assistant at the lighthouse with the duties of companion to the keepers and head rodent catcher—ridding the island of rats and mice. He took his lighthouse keeping duties quite seriously.

Until Sambo came, Nubble was overrun with rats. The

rodents feasted on garbage and food left from picnics tourist had on Cape Neddick. When the rats had their fill on the mainland they invaded the little island and foraged there.

Before Sambo came the lighthouse keepers tried traps, poison bait, brooms, pitchforks, even shotguns to get rid of the rodents. But without much success. But Sambo, with his dog size paws, large jaws, and feline stealth, was a master rat catcher. The cat never rung a fog bell or saved a mariner, but with a heart as fierce as an ocean tempest he protected Nubble and the mainland of Neddick from a rodent infestation.

After completely killing all the rodents on the island, Sambo Tonkus swam the channel to the mainland of Cape Neddick each day, several times a day, stalking rodents in the rocky areas where they hid. The tom cat spent a bit of time carousing with other cats but mostly he hunted mice. Then, he'd swim back to the lighthouse, his prey dangling from his mouth.

Local newspapers ran articles about Sambo and tourists lined the shore waiting to snap pictures of the famous swimming cat.

On Christmas Day 1933, Flying Santa came to the Nubble Lighthouse. Santa flew over as many as ninety-one lighthouses and Coast Guard Stations across New England delivering bundles of toys, candy, coffee, tea, snacks, and shaving supplies.

When Eugene and Amanda Coleman heard the loud roar of a low flying plane passing over their roof, east to west instead of north to south, as usual, they knew it had to be Captain Wincapaw, also known as flying Santa. They dashed to the door to wave at the plane and pick up the bundle he dropped for them.

In 1934, Eugene Coleman wrote a letter describing Sambo's reaction to Santa's visit. *I went out the back door to retrieve the bundle and when I opened the door, I saw Sambo Tonkus (the station cat) flattened out on the ground. The sound of the plane so close had been too much for him and he was scared and glad to get in the house.*

In the 1980's keepers began the tradition of decorating the Nubble lighthouse. Each November, the tower and other buildings are strung with bright, colorful holiday lights. This tradition has also been adopted into a Christmas-in-July event for the summer tourists. These events, as well as stories of the famous swimming cat, Sambo Tonkus, have made Cape Nubble so popular that a photo of the lighthouse was sent into outer space aboard Voyager II, to share it and other information about earth with extraterrestrials.

Billy the Parrot—Portland Head Light, Maine

Joseph Strout, Portland Head's lighthouse keeper from 1904 to 1928 came with an African gray parrot named Billy and a black Labrador retriever named Bo'sun. The parrot would screech commands at the dog, including, "Down, Bo'sun, down!", mimic his bark, and shout,

"Arf! Arf! Sit, you cur!" whenever the dog passed him.

When it was time to turn the fog whistle on the parrot squawked until the boilers were fired up and the horn wailed. Then he rocked on his perch, clucking, "Fog's in; fog's in."

When the keeper's wife's passed Billy's perch or offered her shoulder for him to sit on, he'd say, "Oh, pretty."

If nothing happened around him to trigger one of his phrases then he went with one of his general ones, uttering "Trim sail! All ashore that's going ashore! Or, "Scrock! Bid! Bid! Bid!"

Billy's rambling chatter proved valuable one autumn day when he saved the assistant keeper's son, Don Campbell, from drowning. The boy was told to stay clean because he was dressed in a new suit for his first day of school in Portland. But he forgot, and while his parents were talking with the Strouts, Don found an old toy sailboat, took it to the edge of the slimy rocks, and set it in the waves.

It soon drifted away from shore. The child stretched out one hand as far as he could to catch the boat. In his new leather shoes, his little foot slipped on a slick, moss-covered stone. He helplessly tumbled headfirst into the chill sea. He bobbed to the surface. Not knowing how to swim, the boy flailed desperately.

The tide pulled him under, but Don thrashed his arms wildly, fighting to stay afloat. But only able to keep his head above water for moments at a time, bobbing in the waves. Saltwater burned his throat. Gasping for air between gurgling coughs, his breaking voice could only manage faint, cries of, "Help... Mama... Papa..."

His parents couldn't hear the child's faint, burbling cries.

But Billy, who was perched on Mrs. Strout's shoulder heard the child's dire pleas for help. The parrot flapped his wings and flew to the picket fence above the sea. Seeing the child's dire predicament, Billy alerted everyone by hopping, flapping his wings, and squawking, "Man overboard! Man overboard."

"Whatever is wrong with that old bird?" Mrs. Strout asked, annoyed.

Billy arose from the fence and flew around the two keepers and their wives, screeching, "Man overboard!"

The parrot skimmed Bo'sun's head as he lay sleeping in the sun beside the lighthouse. Billy flew back to the picket fence to resume his flapping and jumping. Bo'sun began barking loudly and urgently.

Keeper Strout could tell something wasn't right. Both animals were spooked. He headed to the fence, breaking into a run when he heard Don's muffled cries. He was able to pull the boy to safety just in time.

Don missed the first day of school, so he could recover from his ordeal with a warm bath, dry clothes, and cup of steaming broth. Perching near the child's side, Billy bent his gray head toward the boy, so Don could scratch it.

As he kept watch over the child that day, the parrot intermittingly screeched, "Man overboard," but in a softer, calmer tone than earlier.

Billy lived a long, happy life at the Portland head lighthouse until Joseph Strout retired.

Then Billy moved in with the Strout's at their new home on a sleepy street in Portland, Maine. And the parrot outlived them both.

This tale of the ever-watchful keepers, and the loyalty, and quick thinking of an unexpected, winged hero, who served alongside them, lives on in the history of the Portland Head Lighthouse.

WE'VE TAKEN a brief look at Lighthouses from Antiquity to the Renaissance, Extraordinary Lighthouse Keepers, Lighthouses in Wartime, Haunted Lighthouses, and Remarkable Lighthouse Pets. There is so much more to the story, but I hope this gives you a taste of the wonder of lighthouses.

That brings us to the afterword of this little book to briefly touch on the state of lighthouses today and in the future.

Afterword

No longer do brave, dedicated individuals trudge up spiral stairs to light lamps or trim wicks for the sake of mariners. Satellites guide even the smallest boats and metallic giants in space silently send signals to ships. The new keepers are remote monitoring systems, automatic mechanisms, solar panels, and sensors.

Many of these sentinels of hope that guided ships through treacherous seas were demolished or left abandoned in the 1960s when countless lighthouses were automated.

However, historic preservation laws fired up a mission to save these beacons. By transforming them into museums, inns, and parks, the legacy and stories of lighthouses and their keepers live on...so, generations to come can step inside their history.

For more information on the lighthouses in this book please take a look at my Bibliography and don't forget to read your complimentary copy of *Specters in the Storm*. After it you'll find a sample of *Scythe2: The Cyborg Chronicles 6*, my sci-fi adventure novel also inspired by lighthouses.

Afterword

[illegible]

SPECTERS IN THE STORM

About Specters in the Storm

Time-twisting paranormal romantic science fiction from USA Today Bestselling Author Pauline Baird Jones

Step into a time-twisting, paranormal whirlwind with Specters in the Storm, the mesmerizing standalone novel by Pauline Baird Jones, known for her smart, quirky, and humorous fiction. This gripping tale of love, science, and otherworldly adventures will leave you breathless.

Dr. Ernest Warren, haunted by the loss of his beloved in the Great Storm of 1901, has renounced love and devoted himself to unraveling the mysteries of science and interstellar travel. When similar, ominous storms begin to brew, he steers his airship towards the past, hoping to unearth answers buried in the place where he lost everything.

Inter-dimensional time traveler, Prudence Pinkerton, has observed Ernest's extraordinary career from the future. Aware of the impending danger, she knows she must meet him and intervene.

Reluctantly, Ernest accepts Prudence's help, setting them on a race against time. As the world teeters on the brink of a life-altering shift, and clocks inexplicably tick faster, they must confront and conquer the evil that lurks in the eye of the storm. But for Ernest, trusting a woman may be an even greater challenge than battling the enigmatic tempest.

Can they defeat the spectral threat hidden in the heart of the storm? Can Ernest open his heart to love again amidst the chaos? Embark on this captivating, time-bending paranormal detective journey in Specters in the Storm.

Purchase Specters in the Storm today to unravel the mysteries hidden within the tempest.

Chapter One

He saw the fog first, then the storm looming above, its dark expanse occasionally cut by flashes of light. The fog rode the storm's wake, billowing and curling like diaphanous bits of feminine clothing. In places the fog wrapped around the storm, long white fingers reaching out, then spiraling back in to cling to the dark mass. In constant motion, the storm itself spun in rotating shades of black and gray, except where the lightning pulsed, like a heart beating in its center. The white fog ebbed and flowed at times, teasing with glimpses of where the storm rode over the strangely calm sea.

As his airship drew closer, the salt-tanged air also brought a hint of metal that left a bitter residue on his tongue. And a hint, no more, of burning wood, teased his nostrils. There must be another ship out of sight, he concluded. If his calculations were correct, they were one to two days from making landfall.

The sight of the sullen storm off the prow of his airship, and the oddness in the air lifted the hairs on his arms and sent his instruments into a frenzy of random readings that made a mockery of his many months of careful observations, not to mention his belief that Science could provide the answers to all questions.

Dr. Ernest Warren lowered his spyglass, handing it to his automaton ship handler to hold while he extracted his log book and noted his observations. All of them, even the ones that offended his sense of logic, were recorded. His task was to observe and report, not to pass judgment on the data he did not like. Later he would attempt to find points of similarity buried within an admittedly incomplete historical record, or in the findings of colleagues. One thing his travels had taught him, the true scientist does not ignore anything, particularly he does not ignore the data that did not fit one's thesis.

He extracted his pocket watch to note the time. But it was as useless to him as his instruments, the hands tracking around the face at close to twice their normal speed. Despite these unusual observations, *The Weatherman* steamed forward with only a moderately increased instability as it rode the air currents. Below him, the waters of the Gulf of Mexico appeared mostly indifferent to the storm, as if sky and water operated separately from one another.

Ernest had not believed the reports that reached him in Europe, had reluctantly turned his airship toward his native soil when the reports became too numerous for a serious weather scientist to ignore. The more lurid of the reports he put down to fear-driven mass hysteria and the remnants of superstition. But when a respected colleague such as Dr. Sebastian Masterson wrote him about the storm, Ernest felt compelled to take note. If this was what his friend had seen, Ernest could see why he was so, he paused his thoughts in search of the right descriptive, disquieted. This storm certainly had many unusual characteristics. But, he reminded himself, serious scientific weather observation was a somewhat modern undertaking. In his studies, he had come across mentions of weather events that he had not personally observed nor had they occurred in his lifetime. That he had never read of a storm such as this one did not mean it had not happened before.

He had the heart and mind of a scientist, but he could concede that Mother Nature was intransigent and precocious. She would lull one into complacency with her sameness, then vary her theme rather violently and abruptly. She did not like to be ignored or taken for granted; hence, the reason she was a Mother and not a Father. Despite this, he still believed that she didn't act without warning, even though it seemed so from their too limited knowledge. He would find the key to her secrets. All of them, so that never again—

His thoughts flinched away from *that* storm. The past that had driven him from his quiet life as an inventor to his quest to finally and completely understand the vagaries of the weather. To defeat its irrationality and its random cruelty, or at least provide a warning when Mother Nature was about to go on the rampage so that adequate precautions could be taken.

He pocketed his notebook and extended his hand for the spyglass. Once secured, directed it at the storm, tracking slowly from side to side, then up, trying to gauge its height. When he could not, he moved his attention to the gap between the storm and the surface of the water. If his eyes did not deceive him, the seas beneath the storm were as calm as those they rode upon.

"Curious," he said aloud.

"Yes, sir," Niles said, with no inflection in his voice. He'd not been programmed for curiosity.

Ernest considered the benefits and risks of a closer look. Though his instrumentation was not behaving, *The Weatherman* itself appeared unaffected by the storm. If it was the storm causing the problems with his instruments, it could be completely unconnected, he reminded himself.

Concluding too quickly closes the mind to further, possibly critically important information. Never assume you know, even when you know. Words of Dr. Masterson to newly anointed weather scientist-to-be, Ernest recalled rather wryly, wishing Masterson were here with him now.

The lightning flashes interested him, and yes, they tempted him. Early testing of his Blizablighter 3000 had been promising but limited by his inability to get beneath enough lightning strikes for sufficient energy collection. He'd rather hoped to encounter a moderate storm during the crossing, but how accurate would the results be in an atypical storm? This storm was bigger than what he'd looked for as well. If his device collected too much charge, could he safely discharge it when water was so highly conductive?

"Let's reduce altitude, Niles," Ernest said, lowering the spyglass. He was most curious about the perceived gap between storm and water. Did his eyes deceive him? The sea liked to play tricks on those who dared to cross it.

"Yes, sir." His automaton used the speaking tube to contact Crispin, his automatonic counterpart, in the engine room.

This change in altitude triggered a threat response in Winston and Fred, bringing them up on deck. Winston still wore the apron of a valet and Fred that of a cook, but those were side functions. They were both higher functioning automatons, primarily programmed to act as research assistants and bodyguards. At the sight of their calm, blank gazes, Ernest found he missed—for the first time—human companionship. Someone who could share the complicated mix of messy emotions roiling his insides. The desire to understand warred with the need to flee this unknown. One might take a vow to be a scientist first and always, but one was, in the end, still human.

"Is there a problem, doctor?" Winston inquired with unruffled calm.

Ernest opened his mouth to explain, but closed it again. They did not have the programming to understand. He did not have that programming. He studied them. "Are you both functioning normally?"

They were also instruments, even if higher functioning than a watch. Both automatons showed the slight eye movements associated with system checks.

"Yes, doctor," Winston said, this echoed by Fred.

Would they know? It was true that the automatons had surfaces resistant to the atmospheric changes and to electrical shocks, but it still seemed odd that they appeared fine, while his other instruments were clearly not.

"Is there a problem, doctor?" This time Fred asked the question, the tenor of his voice pitched just different enough from Winston's so that Ernest could tell them apart.

They had been programmed to be persistent. This worked for Ernest most of the time, but today, less so. He looked at Winston, then Fred. Was there curiosity in the question? Nothing showed in their eyes, but their programming had been designed to adapt and change within certain limits. He'd needed them to learn if they were going to be any help. And they had been of considerable help. They'd saved his life more than once, and they were able to store input for him, both visual and auditory, allowing him access to much more data than his senses could provide.

"This storm may prove to be a problem," he finally said. "It is not behaving...predictably."

Both automatons turned to observe the storm as the airship continued its gradual descent.

"It does have unusual features," Winston said.

Fred nodded agreement, adding, "I am experiencing difficulty in processing the data. It falls outside known parameters."

"I am aware," Ernest said, somewhat dryly. Is this what Mother Nature did to one experiencing hubris? He would not have chosen to return to the Gulf Coast, not now or ever. But it was to this coast that Masterson had asked him to come. Masterson knew his story. He would not have asked if it were not important. The invitation to lecture at the Institute of Meteorological Knowledge in New Orleans had come from Masterson, he suspected. He could not think how else they would have learned of his research. Four years of weather research did not make him an expert by any standard of measurement. Ernest had left these shores, not to satisfy the curiosity of men, but to save lives. He'd been certain that research could and would triumph over dogma and superstition. What price for all his efforts now?

He frowned, considering his options. He'd set this course to avoid Galveston—his brain flinched again—hoping to make landfall at Mobile or Biloxi, then cutting across to New Orleans over land, thus avoiding the locus of his loss. But each time he'd adjusted course to fly around the storm, it had seemed as if it shifted to block his path. It was an illusion, of course. Storms lacked the ability to self direct. For whatever reason, the storm was in his way. He studied the "path" beneath the storm. The fog thickened and thinned, alternatively revealing and hiding the apparently calm seas beneath. It appeared that the gap between water and storm was sufficient for the passage of the airship, though it would leave little maneuvering room for them between sea and sky if something changed unexpectedly.

Was it possible that the storm truly did not interact with the ocean surface? He felt a stirring of scientific excitement, one tempered by the reports he'd perused about the storm. They were troubling and caused him to hesitate. However, the reports he'd read were from land-based encounters. He almost let himself get sidetracked into wondering how it was possible for the storm to appear and disappear in such a wide variety of locations. And why did some people vanish while others were untouched? Such reflections were fruitless and would not get him to his destination. He hated to admit it, but curiosity warred with fear, a wholly natural fear based on learned respect for the forces of nature.

There was one other immutable reality. His fuel reserves. He lacked enough to make it back to Cuba. And if he attempted to fly over the storm and failed, he might not have enough left to attempt passage beneath.

He gave the order, then turned to study the storm again. Would he, he wondered, see the specters in the storm? Or—his personal preference—would he find the proof to dispute what appeared to him to be exaggerated reports.

Chapter Two

IF FRANKLIN PINKERTON had had the foresight, he'd have never named his daughter Prudence. He had known about the family business, however, which seemed to make it an ironic choice—something that Prudence didn't typically associate with her dad as dad or as the Big Daddy of the Agency. Perhaps he'd lost his sense of irony when he took over from his dad who had died in mysterious circumstances—which might make Prudence the hopeful choice, now that she thought about it.

If her dad had hoped she'd be prudent, hope must have died a painful death when Prudence exhibited early signs of the Talent. According to Dr. Masterson, she had it stronger than any Pinkerton since her grandfather. The one who disappeared. At sixteen, Prudence had been flattered and intrigued by finding out she had the jets Talent-wise. At twenty-five, well, it was all Fat City until someone lost an arm during a botched dimensional transit. Or disappeared in mysterious circumstances.

At least she'd made this scene with all her body parts intact, though the transit had been a rough one. The 1905 threads covered the bruising from her dimensional travel, though that was the only good thing about them. Prior to her introduction to the corset, she thought her brassiere was the worst item in her closet.

She should be used to the corsets. She'd lost count of how many versions of 1905 she'd visited while pursuing her Master of Dimensions thesis, which she'd privately dubbed The Ernest Project. With the encouragement of Dr. Masterson, her advisor, she'd postulated the existence of Fixed Inter-Dimensional Iterations and had initially been excited when it appeared that Dr. Ernest Warren fit her construct. Dr. Masterson had cautioned her about the risks of focusing on a single research iteration, even while being equally excited about the research possibilities inherent in a fixed iteration. He'd been right about the risk of getting too close to Ernest. But if she hadn't had her eyeballs on him, she might have missed this aberrant version of Ernest. And the Agency might have missed The Storm until it was too late. If it wasn't already too late.

She patted her nest. Man, the hairdo was almost as bad as the corset, and she needed to get into her bit, inside her head and out, or she'd be popping out the wrong slang at the wrong time. Ernest had the brains to figure her out. She frowned. Had her dad picked this time frame for her study to keep her in the prudent zone?

If he had, it had backfired. Both the storm and the Ernest had stirred up all kinds of excitement and worry within the Agency, though the worry quickly burned down the excitement. Despite years and dimensions of study, inter-dimensional science was filled with lots of theory that was hard to prove or quantify. Each time someone thought they'd proved something, a dimension would pop up disproving that theory.

Like an aberrant Ernest. A deviating iteration.

Her father had wanted to send her brother to check him out, but Dr. Masterson pointed out, rightly, that Prudence knew the terrain, knew Ernest. Masterson's gaze had betrayed serious worry, she recalled, though he had still supported her getting the bit—the assignment, she reminded herself. At the time, she'd thought his worry was all about The Storm, but now she wasn't so sure. She realized that her hands had curled into fists. With some effort she straightened them, flexing them several times to ease the cramping. Had he suspected her lapse in protocol?

She did indeed know Ernest. It was both a gift and a curse. She knew a hundred variations of him, all of them persistently the boss with the jets to rock his time. Of course, he was a nerd in every dimension, either circled or snagged-and-about-to-be-circled by his one true love, Ophelia. She made a gagging sound because she could. That Ophelia kept seeing the cutie beneath the nerd was amazing. Prudence wasn't sure which of them was the more predictable. She could plot their lives on a board. In ink.

Until this one. She should be annoyed he was on track to disprove her thesis.

But everyone was more worried about why the aberrant Ernest and The Storm had appeared around the same time—give or take some math formulas and plus-or-minus cross-dimensional errors.

So that was the gig, er, mission. Find out if the divergence had caused the storm. Or the storm had caused the divergence. Or if they were entirely unrelated.

The divergence. An iteration.

Terms were used to give them distance, to keep them from getting emotionally involved. To keep them from remembering that a divergence was a life-altering event, that the "iteration" was a human being with gray eyes and this straight nose that quivered—

Don't go there, Prudence. It's not...prudent.

Here, at the top of the Sand Island lighthouse, she could admit she'd been—excited instead of annoyed. That she was intrigued by this one Ernest who had taken his rocket off road.

Curious to see for herself what was different—and not just for science and her thesis. Even without The Storm, she'd have found a reason to come. She'd made tracks for Mobile, figuring he'd avoid Galveston, but when the storm blew up, she ran the numbers and decided the lighthouse offered the best chance of intercepting him. Everyone needed a lighthouse in a storm. Was it *a* storm, or The Storm? And how ironic was it that it had come between them? Not unlike the hundreds of Ophelias Ernest had fallen for again and again and again—

But she was over that. She hoped. She couldn't afford to be more than curious. Not now. There was too much at stake, particularly if this was The Storm.

Whichever storm it might be, it was still out of her sight, though her instruments showed it out in the Gulf of Mexico, tracking slowly toward landfall just ahead of Ernest's airship. As the sun began its descent in the west, her view from the lighthouse was tranquil, even soothing, after her rough passage. The lighthouse keeper had been happy for the break, for a chance to take the wife ashore for the night, so Prudence was alone, but she still looked around before she lifted her arm. Her instrument panels were built into decorative leather banding on her sleeves, everything triggered by fingerprint recognition and a cognitive sensor on her pulse. She had to be alert and aware for anything to work. She activated the hologram that would stream the data from the drone she'd launched as soon as she arrived in dimension. She'd collected enough data from his stopover in Cuba that she was confident it was *The Weatherman* she could see steaming steadily forward, now on a direct collision course with the storm. He'd adjusted course twice, possibly attempting to go around the storm.

Both times the storm had appeared as if it shifted to block him.

It was a lot like Ophelia. She bit her lip. In her own time, she'd have had other options for plucking Ernest out of the path of the storm. Her brother would have let the storm have him. Lewis was pragmatic to the tips of his boots. *People get lost at sea all the time,* he would have said. Letting it happen was one way to find out if Ernest and The Storm were connected in some way. But Lewis wasn't here. For some reason, he hadn't been able to transit into this dimension. No one had liked that, but they needed the data, so she got the go ahead.

As the one on the scene, it was her call. She felt in her gut, which she hoped was right and not biased that giving The Storm what it wanted was a bad idea. There was also the risk that removing him from play in this dimension might spawn more dimensions, giving the storm more places from which to draw energy. The truth was, none of them had ever seen a cross-dimensional storm quite like this one, so they were all winging it. What they needed, what they hoped for from her mission, was information. Data was the lifeblood of the Agency.

No one wanted to say it, but this time it might be the saving of the Agency.

Assuming her gut was right, if The Storm did have designs on his person, she only had one option that might get Ernest where she needed him to be. She unlatched the door and scrambled out onto the high balcony that ran around the lighthouse. Yesterday, when she'd come out here, there had been a stiff breeze coming in from the Gulf. Now the air was still and heavy. It smelled different too, a wrongness to it that lifted the hairs on the back of her neck. The menace sensed but not seen, set off a flight-or-fight response, and she had to take several deep breaths to close down the dimensional portal that tried to form in front of her.

When the portal faded to a faint shimmer, she looked seaward. Could she see clouds on the distant horizon, or did she see them because she knew a storm was there? She watched *The Weatherman* dot for a couple of seconds more. Adjusted the angle. Did it seem as if it had reduced altitude? She'd heard of ships flying over a storm, but under? She shut down the holo and switched over to communications, took a deep breath, and keyed in the code that would connect it with its counterpart on her familiar.

"Octavius? Baby, go fetch." She launched the pre-loaded data that would help him home in on *The Weatherman.*

Out in the waves just off the shore, the water boiled as Octavius surfaced just long enough to wave a tentacle at her. She waved back, watching the rounded dome sink out of sight, then a wake appeared, the crested "V" pointing straight out to sea. To Ernest. If the octopus made it past this storm.

As far as they knew, The Storm had shown no interest in nonhumans. As far as they knew.

They knew so little. She prayed for his safe passage. If she lost Octavius—she shook off the fear as a portal tried to form again. Pulled up her holo-tracking. All the data they had indicated that The Storm spun above both land and sea, the gap both inexplicable and puzzling. It had to touch down sometimes, though, since people had disappeared in the places it had appeared. No one understood why The Storm took this person, but not that one. Or why so few land grabs. Why it preferred the sea and those who sailed it. There seemed to be no pattern to where it appeared, or why it disappeared just as abruptly. Their difference engine computers had tried every combination of the data collected. Even the so-called link to this dimension, to this Ernest, was a tenuous one.

The stories of what those not taken had seen, well, none of their kind could confirm or disprove the specters in the storm. Or even theorize what they could mean. Her brother had volunteered to ride out The Storm when it popped up in their home dimension, but it had vanished before he could get under it. That left this dimension, with this aberrant Ernest, as their only hope. No one was certain there was a connection. Or that there wasn't. But it was the only dimension not showing stress from The Storm. If this was The Storm, she reminded herself.

It wasn't proof positive, this storm shifting to block his path, but it was troubling.

An atypical Ernest and now an atypical storm.

Was it stalking Ernest? Or worse, was his aberrant iteration the cause of The Storm, as her father and others believed?

Chapter Three

ERNEST DIDN'T POSITION Winston at the prow, though he'd been tempted. It was their function to occupy high-risk places, but he needed human eyes and instincts, his human responses, as they cautiously chugged toward the edge of the storm. Below them, the sea appeared to grow calmer—almost becalmed—as they pierced the first edges of the trailing fog. Wisps of it wrapped around their stacks and trailed along the sides of the airship like very long, exploratory fingers.

There was a small bump as if they'd passed through something resistant, then a jerk as the airship's speed increased slightly. It almost seemed, after that initial resistance, that something now pulled them forward faster. Alarmed by that thought, he rapped out an order, and the engines slowed. To his relief, so did their forward speed. Conversely, he now wished he hadn't slowed. A pure scientist, he had not believed in anything outside his sight and sound, but he'd learned the hard way that there were other senses, ones that warned of danger before it was seen. If he'd trusted it back then—

He snapped off the thought that threatened to pull him into the past, distracting him from this very present peril.

The only sound came from the engines and those sounded muffled, as though the thick air acted as a dampener. Even the sound of his breathing seemed muted, felt labored as if the thick air entered his lungs reluctantly.

When they were fully under the storm, the fog fell away, drifting back toward the outer edge once more. There were probably air currents he could not feel, he told himself, currents caused by their passage. He needed the rational in the face of the irrational. It was darker under the storm, but it would be. Though the water ahead was calm, the storm showed an observable circulation, at least from what he could see through the murk. The shape of the storm was that of something viscous pushed through a funnel, the spirals thicker at the edges, then thinning as they approached the center. He was too far from that center to see if there was a defined eye, though this storm appeared to lack other recorded characteristics of a cyclonic storm.

The smell of burning was stronger now, and the air tasted coppery. He was reminded, he frowned, of a blacksmith's forge. Fire and brimstone. It did not fit the general conception of hell. Fire and water were, for the most part, mutually exclusive. Unless there was something in there...his skin crawled, and he felt an urgent need to turn back. He suppressed it with an effort.

The stacks shuddered as if they'd come into contact with something. The top of the dirigible was lost in the dark clouds, so he ordered a few more feet of descent. The sturdy airship drifted lower, skimming along just barely above the surface of the almost motionless water. The shudder stopped.

They moved forward with more ease. It became easier to breathe. Once more he resisted the urge to increase speed. It would be unwise to expend too much of their fuel reserves when there was so much he didn't know about what might be ahead. Their progress was steady and unimpeded. He should have felt better.

He didn't.

The sensation of being watched crept over him. He needed to look up. He feared to do so. Off to his left, the setting sun sent rays of light under the storm, though this light did not reach them. It acted more like a distant beacon. And indicated that the storm had limits, an end if only he could reach it. The sight of the distant light eased the growing panic, gave him hope, though the sense of being watched did not ease. If anything, it became worse. The sensation was one of animus, but more than that. He felt fear in there. Anger, malice, nothing positive. His hands gripped the gunwale until his knuckles turned white. He tried to speak, but his throat went dry and closed as if something gripped it to hold in the call for assistance from his automatons.

He wanted to claw at that hold, but he couldn't let go of the gunwale. Couldn't move. His chin started to lift as if impelled. When he could fight no longer, when he thought he must look and die, he was distracted by movement along the surface of the water. He blinked, sure it must be an illusion. But the sight of it, for whatever reason, eased the sense that he could not move or speak.

"Winston! 'Ware!" He called out the alert command. This was not his imagination, he realized. It was real. The wake made a perfect "V," like an arrow pointing straight at him, with waves falling away from the edges of something just under the surface. It came on, swift and straight as an arrow.

A sort of howl, like the wind, but not like it either, sounded above him, then the surface of the water came to a boil. Out of the maelstrom, he saw—

Tentacles?

A different kind of panic rose inside him. Giant tentacles.

He started to back from the edge as two of the monstrous things reached up, sliding along the bow, then gripping the gunwale close to where he'd stood.

"Ware!" he called again, the sound more a croak than a call.

A large red dome rose from the water and giant eyes regarded him for a long moment.

There was another howl, like a wind he couldn't feel. A jerk. Then a jolt as *The Weatherman* slammed into the surface of the water, yanked downward by the tentacled grip.

Ernest flew backwards hitting the deck hard enough to knock the wind out of him. Just before his head connected with wood, he saw what he'd feared to see...

...faces...

Thousands of them.

The specters in the storm...

His eyes widened as he saw one he knew better than his own. She drifted just out of his reach. There was another wail, but this one vibrated inside his head as if he had made the sound. He reached out, but his head hit wood, and the darkness closed in, taking her away from him again.

Chapter Four

PRUDENCE WAITED on the lighthouse's balcony until her tracking screen showed contact between Octavius and *The Weatherman.* Now together, both raced toward her position. Octavius could haul tentacles when he had to. Or when he was scared. Anxious for him, she went down the winding stairs and outside where tumbled granite blocks surrounded the lighthouse. She glanced back at the light keeper's cottage. Prudence had left a lit lantern on the porch, certain she'd need the light by the time Octavius returned with *The Weatherman.*

The tide was out, she noted. Usually, there would be seagulls checking out the small pools left between the chunks of granite by the retreating water, but all was quiet as a grave. She picked her way as far as she could across the large chunks. Now she could see the water line where small waves rippled ashore, much smaller than normal. Of course, what was normal for the twenty-four hours she'd been in the area? She'd done a bit of cramming before her transit but didn't know then what mattered and what didn't. Or that she'd remember the right things. This Ernest had caused her all kinds of headaches. She should be angry, not—

She checked tracking. Octavius was still within the boundaries of the storm, but moving at his top speed. The setting sun was almost parallel with the water now, sending beams of gold and orange across the almost still surface.

If they didn't show up soon, she'd have to go light the beacon. Another crash course, this time from the light keeper. Hoped it went well.

Octavius had the ship, but did he have the passengers? The storm had not changed speed, at least not enough to register on her equipment. She was about to turn back when she saw it.

A dark speck that rapidly grew larger against the horizon.

It resolved into the shape of a ship skimming along the surface of the water, and just ahead of it, a hint of a rounded red dome. Red. Something had made Octavius angry. At least he wasn't scared. One of them shouldn't be.

There was still enough light for her to see a figure at the helm, but there was a stillness to it that hinted it was an automaton in power conservation mode. Octavius reduced speed, churning up a counter wake. He released the airship, and it drifted towards her, just above the water. That brought the automaton back to attention. He steered it gently around the rock outcrop, and then toward the pier. Prudence followed it, taking the rope when tossed and securing it. The dirigible's balloon kept the airship above the water when the engine cut off. It bumped against the edge of the pier.

Octavius drifted in close and waved a tentacle at her. She was relieved to see his natural color had returned. She waved, said into her wrist, "Thank you, sweet baby. Good job."

One tentacle touched his rounded dome, as if in a salute, then he sank into the water until just his eyes and round dome were visible. Beyond him, and not her imagination this time, the horizon darkened. Whatever it was, a storm or the Storm, it was incoming.

"Be careful out there," she added, lowering her arm once more. It was a pity one couldn't debrief an octopus. He'd been in the belly of the beast. And he'd turned red in anger, not white in fear. That meant something, but she wasn't sure what. She sighed and turned toward the airship. If she turned colors like her octopus, she would be white right now. What Ernest would she find onboard? If she found one. She studied his airship. Looked like a sweet ride for an airship. Ernest, being Ernest, had probably made some modifications. His iterations had done some interesting things to his steam car. She wished she could check it out.

Perhaps later, if there was time. It was one of the ironies of their kind that they could travel through dimensional time, as long as it was not their own, and yet still be limited *by* time.

From her vantage, she could see along the length of this side of the airship, but a section of the stern of the ship was obscured by the main cabin. She saw a booted foot poking out from behind that cabin, a slackness to it that was a worry. The gunwale was low enough for boarding even with the multitude of skirts, so she gathered them up and climbed aboard, proceeding cautiously toward the foot. She cleared the cabin and stopped. Two automatons crouched protectively next to the prone figure. She didn't have time to study him, not with a couple of alert automatons watching him, but there was no question he was *an* Ernest.

The rise and fall of his chest was a relief. It wasn't a surprise Octavius had delivered Dr. Warren in an unconscious condition. Most people reacted to the sight of the giant octopus by fainting, though, based on the traces of blood on the deck, Ernest had been knocked out before he could pass out. That might help. Men hated to faint. Not that they loved getting knocked out. Or giant octopuses. She wasn't too worried because there was so much more to come that he wouldn't like.

She stopped, opting to give the automatonic guardians time to assess her threat level. Thankfully they didn't have that ability, at least not to do it accurately. Prudence knew how to look harmless and helpful. Not to mention how to do helpless 1905 lady. She gave a dramatic start and covered her mouth with the tips of her fingers.

"Oh." She widened her eyes and made a couple of ineffective hand gestures, learned from watching Ernest's vapid fiancé through far too many dimensions. "Is he all right? Oh, my goodness sakes alive." She fluttered her lashes.

"He is unconscious," one of them finally said.

Prudence pressed a hand to her chest now. "Gracious! Should you leave him there?" She gestured toward the shore. "Would you like to bring him inside the cottage? There is a bed you can use." The quarters were spartan, but they could have been worse if she'd had to time her arrival, say, a year from now when a hurricane would take out the cottage.

The automatons would not find it odd to find her alone and in charge of a lighthouse, or at least they shouldn't. What programmer of a difference engine would waste precious storage on unlikely scenarios like that? Asking the question, though, instead of just ordering them, would give her a sense of how complex their programming was. The automaton who'd been at the wheel stumped to the edge and looked down on them from the wheelhouse. Another one came up from what she assumed was the engine room. The situation clearly exceeded their programming. They visibly stalled in place, their eye movement tracking up and down.

The other two showed signs of "thinking," their metal eyes moving from side to side.

"There is a storm incoming," she added, hoping to tip the balance to cooperation. If their programming perceived her as an ally, everything would be easier, well, less hard, she amended. She'd left easy in her own dimension when her Talent manifested.

The mention of the storm appeared to be the right nudge. One moved to Ernest's head and lifted him by the shoulders, while the other secured him about the knees.

Bodyguard automatons. Normally their programming would

make them want to keep him on the airship, but recent events, an octopus—still visible out in the water, its eyes appearing and disappearing as it rode the small waves rippling ashore—combined with the storm threat, would alter that imperative, sending them seeking the closest secure structure. The lighthouse was the sturdiest thing around, but the cottage had better amenities. As she turned, she gave Octavius a quick wink and another out-of-sight wave.

The light was fading fast, the lantern she'd left on the cottage's porch a cheerful beacon. She navigated the gunwale, certain the automatons wouldn't care about proprieties. She stepped back so they could clamber off, casting a last look of regret at the airship. The two ship handlers did not attempt to follow them, though they might become active if she came back without Ernest. He would have needed to secure his retreat during his travels. This world, this time in its history, had many treacherous places. Had this Ernest been seeking redemption or a final solution during his travels, she wondered?

The two automatons headed for the light, following the narrow boardwalks, first taking the one that led to the lighthouse. She followed them, finally letting herself study this Ernest as the last of the light fell on him, let herself look at the one Ernest whose life had not followed the pattern of all the others. His face was very tanned, not a surprise since he'd been roaming around studying the weather. Otherwise, he looked the same. The face was thin and narrow, with a high, intelligent forehead and a grimness about the mouth, apparent even in his unconscious state. It was, she knew now, his look of repose. And when he thought. Which he did too much, except around Ophelia when he thought too little—

She clamped down on this. His dark hair was untidy and in need of a trim. He also needed a shave. This was new. She'd never seen any of the Ernests anything but meticulously turned out. He was a scientist, but not of the crazy or absentminded bent. He applied the same attention to his attire as he did to his studies. Of course, he had been knocked on his backside by an octopus.

His frame was long and somewhat leaner than the others, his travel clothing more tidy than his hair, despite his tumble. His eyes would be gray under the lids. When they lifted, what expression would this Ernest's eyes have?

She'd known the others felt it. It wasn't just her. At first, she'd been shocked and retreated because of *rules*. But then—she'd started looking for it. That moment when he realized that he could want more. His pupils had dilated. His pulse had quickened. There'd been other signs, which had embarrassed and troubled the oh-so-loyal Ernests. Because none of them could stand to be wrong, they'd opted for manufactured dislike. She'd let them push her away. One couldn't fight for what one couldn't have. So she'd faded from their lives, wondering if any or all of them ever thought of her after she was gone...

But this one...

She kept a tight hold on hope because she was not only out of his time but also out of his dimension. That didn't mean she couldn't wonder what might be different about an Ernest who had loved and lost. One who couldn't have Ophelia. One who had had four years to heal...

The automatons turned onto the boardwalk that led to the cottage.

"Take him inside. You'll see the bed," she directed. "I need to light the beacon." She had to do it. She'd promised the light keeper. She just hoped she wasn't painting a target on them for the storm.

Chapter Five

WHEN SHE'D LIT the beacon, Prudence made her way to the cottage. The two automatons had stationed themselves against the wall on either side of the head of the bed, their mechanical gazes turning toward her as she entered. Since they could not show emotion, their blank gazes gave no indication of what their next moves would be.

"How badly is he injured?" Automatons liked questions—if they could like anything.

"He hit his cranium," one of them said.

"He has been unconscious since the attack," the other added.

Attack? She preferred to think of it as a rescue. Interesting that, while they looked identical, the automatons' voices had been programmed to sound different.

"Do you have designations?"

"I am Fred."

"I am Winston."

Names, not numbers. This was another divergence from the other Ernests. This Ernest had fled his grief and mostly eschewed human contact—but had given his automatons names. In all the other dimensions, he was at home in Galveston with his wife. All of those Ernests and Ophelias had survived the storm of 1900. Only in this dimension was it the Great Storm of 1900, resulting in a devastating loss of life, including that of his fiancé and her father. The loss of his Ophelia had spawned more differences in Ernest's observable life, but so far these had not resulted in the formation of new dimensions. This went against known dimensional science. And in an interesting dichotomy, there was The Storm, which had appeared across all dimensions—though not in the same places in each—which was again not typical of what was known. Or believed to be known to a science even more imprecise than she'd realized.

This was the first dimension where the storm had appeared in the Gulf. And it was only in this dimension that Ernest had left Galveston. If the storm wanted an Ernest, it could have easily found one of the predictable Ernests. Instead, it appeared to want this one. He'd traveled the world for the last four years. The storm had also cruised around. Had their paths crossed at some point? Or not crossed, but intersected? Was there a connection between this aberrant Ernest and the storm? If that was The Storm heading toward them. It could just be *a* storm.

That was the question she'd been sent to answer. The hope being that in the answer lay the way to stop The Storm. Because it didn't just appear in all dimensions. The Storm seemed to be drawing energy from the other dimensions, degrading and destabilizing them, eating at the edges like acid. What was different about this dimension? Or was she looking at that difference? Or was the difference something he'd learned in his travels?

She took a deep breath and looked at the automatons. "With the storm coming, I'm sure Dr. Warren would like his gear and instruments, all his research material, secured from the airship." She gave them time to process this information, then added, "The safest structure in the storm is probably the lighthouse. Stow it in the area below the lamp. That will be the safest place."

Prudence wasn't sure about up being safe if it was The Storm.

It might put them up in the very heart of it. But she hadn't been sent here to do the safe thing. And the lighthouse walls were some thick. That should help. Because she hadn't been sent to die either.

She looked at the one called Fred. "Perhaps you could stay with the doctor and Winston could secure the ship and the doctor's things? I'm sure his ship handlers would help?" She'd pitched her tone to be authoritative, but hopefully not too much. Bodyguard automatons were tricky.

After a processing pause, Winston nodded and left the room. Prudence waited until he'd thumped out of the cottage and then said, "I'll get some water for his injury."

ERNEST WOKE with an aching head and a significant gap in his memory. He'd been on *The Weatherman,* and now he wasn't. How had that happened? He kept his lashes down and studied the woman from under them, not ready to confess he was awake until he felt more certain what he was waking to.

So far there was just the woman, but that did not help with his unease.

He avoided women. Women reminded him of what he had lost. Or worse, tried to console him. Tried to fill the empty place. It was best to just avoid them as much as possible. Only this time, he couldn't get away, couldn't avoid.

There was something about this one that bothered him more than usual.

She was attractive, he acknowledged with extreme reluctance, felt an immediate stab of guilt at this disloyalty. His friends told him he couldn't, shouldn't mourn forever, but he'd lived while she died. It should have been the other way around. She'd been so bright, so full of life that he couldn't believe she'd agreed to marry him, a dry, dull inventor, over all the other more interesting suitors who'd vied for her hand. But she had. He'd sensed the signs of a possibly dangerous storm, despite the lack of concern by those who should have known. He'd shared his concerns, had asked her to leave with him. She had listened to her father and the other "experts." He tried to find her in it, but the storm won. It beat him back. Kept him from doing his duty. She and her father had died. All he could do for her now was remain loyal. Never forget.

He wouldn't have looked, noticed this woman if he'd not been at such a disadvantage, he insisted to himself. Any man, alone in a bedroom with a strange woman would be unsettled and feel caught off guard. She looked at him like she knew he watched her. There was both sympathy and humor and something else in her eyes. In an effort at regaining control, he lifted his lashes, trying for dispassionate distance, to see her as a scientist might. What was she doing here? He looked around. Where was here?

The room was spartan, spare. Some effort had been made to personalize it. A few photos. Some of those things women made for the tops of dressers. They were close to the sea, based on the humidity in the air and the faint smell of the ocean.

Because he could not help it, his gaze wandered back to her. Brown eyes and hair, no way to assess her height while she sat. She was dressed for travel. Her corset was leather and durable, her boots sturdy and sensible. She'd left off her hat, but goggles hung about her neck. Her sleeves below the elbow sported intricate brass and leatherwork, the panels running down the top of her arms. An odd conceit in an otherwise practical ensemble.

Why did he sense impatience, restlessness when she sat with outward stillness, her hands folded, unmoving. A slight tilt to her head presented a jawline that was clean in its lines with a hint of stubbornness. There was intelligence in her eyes and humor softened the line of her mouth.

His gaze lingered on her mouth until it curved up.

He jerked, his hands clenching on the coarse blanket. A gentleman did not—

"Does your head hurt very badly?"

Her voice was clear and cool, like the mountain streams after a storm—he pulled himself into a semi-sitting position, for the first time becoming aware of Henry to one side. The automaton moved to assist him to sit, tucking the pillow behind his back before retreating again.

Pieces of memory began to return as if sitting up had shaken them loose in his head. The storm. There'd been a strange storm. Had he dreamt that? He must have because there'd also been a giant octopus—

"I'm sorry that Octavius startled you. He's young, and he got a bit excited, a little over enthusiastic. He's still learning to manage all those arms." She shrugged and moved her arms like tentacles.

"I didn't dream it." He didn't know whether to be relieved or angry or something in between.

Her elegant brows rose. "No. Octavius is quite real, though not at all malicious. He was the best available option for getting you clear of the storm, you see, or I wouldn't have risked his life."

The storm. The specters in the storm. Had he really seen Ophelia just before—

"I promised Dr. Masterson that I'd try to help make sure you made it safely ashore." Her mouth curved up a bit ironically. "I should be more careful what I promise."

Perhaps he had not yet waked from his strange dream. Light filtered down from above, as well from a lantern. He could smell coffee and food of some kind. If this was a dream, it was most vivid.

"Where am I?"

"This is the Sand Island, off Mobile Bay, and we're in the light keeper's cottage."

He processed this for a few seconds. Then he frowned. "How long was I unconscious?"

She consulted a watch extracted from a pocket on her leather corset vest. Gave it a slight shake and restored it to its pocket. "I did not note the time, but you've been out an hour since you arrived here."

His frown deepened. "But...I was some distance from making landfall." He no longer felt sure of his calculations. He looked at his automaton. Before he could ask, she spoke again.

"Octavius is a very fast swimmer. Octopuses are, you know. They can't swim fast for long periods, but he didn't like the storm." Something in her eyes changed. She shifted in her chair. "What was it like?"

He did not want to talk about that. And he felt acutely the disadvantage of being prone, even if she was seated. This—none of it made sense. More than he needed for his head to stop aching, he needed to reach a place where something made sense. Anything. He shifted his legs to the edge and lowered them to the wooden floor. It helped to get them on something solid, though his head swam unpleasantly. He rubbed his face. It did not help, so he looked at her again.

"I do not wish to be rude, but who are you?"

Her eyes widened. "Oh, I'm sorry. I forgot you haven't met me."

Why did her words seem to imply she'd met him? He blinked. That was not possible. And yet...there was something familiar about her.

"I'm Prudence Pinkerton."

He'd heard the name Pinkerton, heard of the detective agency, but she was a woman, a female. And if she were a detective, what would she detect here? Not that she'd claimed to be one. She'd asked about the storm, but a storm would not be of interest to a detective agency. Vaguely he knew the Pinkertons dealt with crimes, such as train robberies and the like, or so he'd heard. Why would Masterson have dealings with them? Or have sent this woman to help him make it ashore? Masterson could not have known about the storm.

He felt slow and stupid, feelings he did not like. Men were not slow and stupid in the presence of women. They were—he slammed his thoughts down. He was through with women. Masterson knew this, and still he'd sent this one. A woman with a giant octopus. Did Masterson know about that? He cast her a look of suspicion.

Her lips twitched as if she read his thoughts. Ophelia used to call him a closed book, but this one appeared to read him easily enough.

"I could tell you how I came to have Octavius, but it wouldn't help."

She seemed oddly comfortable with being alone with him. She appeared respectable, but there was also something not...not that she looked like a loose woman, it wasn't that. He studied her. Her ease was more of that between two men, as if they were equal. Was she a suffragette? He avoided those as well. Where had they met?

"You are," he paused, adjusting his assessment to something more polite, "most unusual."

She grinned. "Yes, I am." Her head tilted to one side as she gave him an equally direct survey. "So are you, Dr. Warren." She paused, then added, "I've never met anyone being stalked by a storm."

He stiffened. "It is most atypical, but—" He stopped when her brows rose over an abruptly ironic gaze. His mouth set in a stubborn line. "Storms do not stalk people. While it may appear to have been on the same course, there will be a logical explanation for it."

She rested an elbow on the arm of the rocker, then rested her chin in the palm of her hand. "And what is that logical explanation, Dr. Warren?"

He opened his mouth. Closed it. Finally muttered, "If it is following me, which I doubt very much—"

"It changed course after Octavius collected you. It's following a straight line that will bring it to this location by morning if it maintains its current speed. Sooner if it speeds up."

He blinked, trying to figure out which part of this was the most unbelievable. "How could you know its current speed or course? Did your octopus tell you?"

She looked amused. "No, Octavius does many useful things, but that isn't one of them." She hesitated. "I could tell you how I know, but you'd have to believe six impossible things before breakfast. Can you do that, Doctor?"

Chapter Six

PRUDENCE HAD BEEN sure she'd *know*, once she saw Ernest, once she looked into his eyes, and she did know one thing. He had not consciously caused the storm. He was not evil or even misguided. He was bewildered. What she didn't know was if he was the reason for the storm, which was a different way of being responsible. He could still have done something to trigger it. Or this dimension could be the problem, and he was a side issue.

Did she need to kill him or recruit him? Could she execute the Final Option? Her heart said no. Her brain said she had to if—but didn't think she could do it either. She wasn't a killer. They all knew that, and they'd let her go, so they didn't think she'd have to either, she reminded herself. She knew what she wanted to do, but what if what she wanted to be true blinded her to what really was?

She should have told her father that she wasn't unbiased enough for this mission, but if she had, she wouldn't have just lost the mission. She'd have been out of the Agency. Do not get emotionally involved with dimensional iterations was one of the rules an agent couldn't break and remain an agent, the line that could not be crossed. Ever.

Someone should have told her why the line couldn't be crossed. It might have made a difference. Maybe. Knowing how much it could hurt might have headed her off. It was human nature to risk pain, to pursue the unreachable, to dream big and try harder. Being an agent, required all of those things. She'd broken bones learning to navigate dimensions. It hadn't stopped her. A broken heart, well, that was a different kind of pain, particularly in a hopeless cause. If she'd known, she'd have tried not to let it happen, wouldn't she?

Only she had let it happen. And she'd come to protect Ernest. She knew it, even if she didn't want to admit it. Not because she thought that this time she'd get him, but because she couldn't let him die unless—but what if she had to? He was one Ernest in thousands—only, he wasn't. He was an aberrant Ernest. And therefore unique. Special.

Against special, there was only the fate of innumerable dimensions. Billions of lives, including those of her family, depending on her to make the right decision. To do the right thing.

No pressure.

She needed more information, but it wouldn't be easy. This Ernest was like the others in one essential way.

He was a scientist. He probably couldn't believe one impossible thing—let alone six—if she had a month of mornings to spend. Which she didn't.

She had one night to get him from skeptical to helpful. And she could only think of one way to make that happen.

Break the other unbreakable rule.

Never tell an iteration they are an iteration or that dimensions could be crossed.

Since she was already in for a pound, why wasn't she spilling the beans?

Because a part of her, a very small part of her, wasn't sure she was right. If Ernest was a bad guy very good at pretending to be good and she told him? Well, it wouldn't just be bad.

It would be catastrophic.

Six impossible things. Was this what it was like through the looking glass? There was a small one tacked to the wall, but Ernest didn't go to it when he clambered to his feet. Instead, he made his way, somewhat shakily, to the slightly opened window and looked out. An unfamiliar, diffuse light revealed chunky blocks of stone and sea. From here it appeared that the house stood in the sea, was surrounded by it. As unsettling as that was, what still struck him the most was the silence. He should have heard something. The sound of gulls. The slap of water against rock. Something. Here was one impossible thing he had to believe, whether he wanted to or not. And it was true that the storm behaved like no storm he'd observed. Or read about. He turned from the window, from the silence, but not toward her.

Losing Ophelia had not been the impossible. People died. It was a fact of life. He'd seen it before. He'd seen it after. Seeing her in that storm—his hands clenched at his sides. It had been Ophelia, but not the way he remembered her. Her eyes when alive had been, well, he could not recall exactly. But not like the eyes in the storm. Those eyes, that face...had not been...

A faint shudder shook him.

"What did you see in the storm, Doctor Warren?" her voice was soft, but something in it, an urgency perhaps, made him turn around and face her.

He opened his mouth, but could not say the words.

"You saw her, didn't you?"

After a pause, he gave a short, sharp nod. "How did you know?"

"You're not the only one to see the specter of a...loved one in the storm." Her gaze shifted away from his, but not before he caught sight of sadness in them that he recognized.

"You have lost someone."

She nodded.

"In the storm?"

She shook her head. "I've read witness accounts." She looked past him, in the direction of the storm. "This will be my first contact with it." Her gaze shifted seaward. "If that is The Storm."

"Why are you here?" Why had Masterson sent her? Why had he sent a woman? And what did she—or he—expect to happen?

She didn't speak for so long, he thought she wouldn't. Finally, she stirred, her lips curving wryly up. She looked at him once more. "To stop the storm."

Her tone said it was obvious. His brows shot up, even though some part of him was, strangely, not surprised. "That would be one of the impossible things I am supposed to believe?"

She smiled, transforming her face in a such a way that it caused his heart to thump oddly.

"Only five more to go."

Chapter Seven

Ernest was recovering. He was thinking. That could be good. That could be bad.

He was like the other Ernests, but also different. His wanderings in his airship had altered him. She'd like to think he'd grown, matured, become less rigid. She wasn't sure, though. She might be, not just hopeful, but dangerously hopeful.

She still found it hard to believe that all those Ernests had loved all those Ophelias. In an effort at impartiality, she'd spend some time with some of them, trying to see what he saw. That had only made it harder. Some were not so bad, but none were what she'd call worthy of great devotion. Most had started to make their Ernest unhappy on the heels of their wedding vows. The Ophelias weren't outright evil, and they did show some sense in wanting Ernest, even if he was totally wrong for them. The main problem was they couldn't leave well enough alone. They wanted things for the Ernests that the Ernests didn't want for themselves. Like wanting him to work for her daddy and give up that crazy inventing. They couldn't see what he was because they were too focused on what they wanted him to be. It should have helped her get over Ernest because he'd done the proposing, but to be fair, how did a man of his time—or all the times when women were limited so severely—find spouses who were a match for them? Even in her time men—even smart men—fell for the pretty face without looking any deeper and sometimes regretted it. Within the Agency she had sometimes felt the sting of being judged as a girl instead of for her Talent, though not as much as at school. All that had mattered in high school was her poodle skirt and who thought she was a cutie. And the guy who thought she was a cutie had to have a souped up dragster or having a guy wasn't enough to make one cool. Inside the Agency, she had a future, but without it? Her guidance counselor had urged her toward nurse or school teacher or secretary. If he could see her now? He'd probably cry like a girl.

Though to be fair to him, it also bothered him when she rolled up to his car on her skates. He had wanted better for her than a drive-in job and couldn't know she needed a for-show job that let her do her real job.

Honestly, she didn't know if she was a cutie or a dog. She'd been told her looked like her grandmother, the one married to the missing grandfather. She'd never met either. Didn't want to cause a paradox. Those didn't get you kicked out because most likely you'd erase yourself from history before they could kick you out. The problem with popping through dimensions, you didn't get a lot of time to date. And if one made the mistake of getting a crush on an iteration? Yeah, that tended to kill your love life, too.

She knew she didn't feel comfortable with the idea of seeing herself through someone else's eyes, letting that dictate how she saw herself. So that put her out of step with her peers, too. Even though she pined for Ernest, she didn't *pine* to the point of despair. She tried to keep it wistful, though she had to fight a bit of melancholy after a mission. But she'd always known she couldn't have him. If you started out with no hope, the only way to go was up. At least in theory. But she couldn't remember ever wondering what was wrong with her because it was obvious the problem lay with Ernest. And guys in general who forgot to use the correct brain when choosing a life's companion.

There was a clink as Ernest set down his utensils and the scrape as he pushed his chair back from the rough hewn table. He'd agreed to eat, perhaps hoping she'd be a better cook than his automaton crew. Oh well. He'd eaten. She could have done better in her own time, of course. She'd taken Home Ec. But it didn't prepare one for a wood stove.

He studied her now, with his scientist look back on his face. She'd seen it before this meeting, over and over again on the faces of the other Ernests. She held the look without flinching or blinking. She could have held it in her sleep, except then her eyes would be closed. He looked away first, though his gaze quickly returned.

"No one can stop a storm," he said flatly, as if he'd been thinking while he ate and drank. Which he probably had. All the Ernest Warrens were brilliant scientists, ahead of their time, which made the Ophelias not just wrong, but blindly stupidly wrong. All of them but this one had directed their brilliance toward inventing things, though all of them had allowed the inventing to become after work hobbies. Only this one had turned his attention to the elements, but he still invented things, according to Dr. Masterson. This iteration, this Ernest, focused on inventing weather instruments that he hoped would help him figure out the weather. Was that a clue to the storm's interest? Maybe he'd invented a way to stop it, but didn't know it?

"This is not a normal storm," she pointed out, her tone mild. She rose and began clearing the table, partly because it was expected and partly to avoid that look of distrust in his eyes. She set some dishes in the big farm sink and turned around.

"You are saying this storm is man made?" He could not keep the disbelief out of his voice.

"I did not. I don't know what's causing the storm," she added before he could ask. She crossed her arms over her chest. "You've seen it up close. Is it normal?"

He glared at her, then looked away again. "It is atypical," he conceded some ground, but not all of it since he used the word "atypical" again.

"It is unnatural," she asserted because he needed to get that far if they were going to make any progress toward those other impossible things he needed to believe.

"The study of weather—"

"Science may not be ready to admit this is unnatural, doctor, but you know it with your other sense. Your instinct." She rested her hands on the table and leaned in. Could she say gut in this time? The need hadn't come up before. He should have good instincts, after wandering around the world for four years. "Your mind may dispute this because minds don't like the unnatural, but you *know* in your gut and your heart what it is." Did he still have a heart? Was it still broken? The distrust in his eyes was more than the others, but this one had a better reason for that distrust.

His mouth straightened into a thin, tight line. But he gave a short, sharp nod that might have been a concession.

"But if you don't know what is causing it, then how do you propose to stop it?" He didn't try to soften his scorn.

"Ah." He'd put a finger on the painful crux of the matter. Stalling for time, she returned his serve. "You're the weatherman."

"Is this another of your impossible things? I have just begun my research on weather phenomenon. A mystery such as this one is well beyond my skills."

"Maybe, but the storm doesn't seem to know that."

"What?" The single word was snapped. "What are you talking about?"

"It's following you. Either you're attracting it—" she paused, with the sense that she was close to knowing something, just not close enough to know what she knew...

"Or?"

"Or you have something it wants." She frowned. What could a storm want?

"What could a storm want?" He echoed her thought, but with even more scorn than before.

"That's a very good question." She said it slowly, her thoughts spinning, but with a stronger feeling of something important just out of her reach. She sighed. "Let's start with this."

She reached for the map she'd left leaning against the table leg, pushing aside the cutlery and dishes she hadn't cleared, so that she could spread it out on the surface of the table. Against his will, interest sparked in his eyes. For a map. She sighed a little inside.

"What is this?" He bent over it, studying it as if he couldn't wait for her answer. He must figure it out for himself. His turn to frown. "It's a map of the world. A most detailed map."

Definitely better than anything he had access to, but hindsight was twenty-twenty.

"What are these marks?" he asked.

"Those are storm sightings."

"I didn't realize there'd been so many."

There hadn't been. In this dimension. But he wasn't ready to hear that.

He looked at it without speaking for several minutes, then frowned, his gaze narrowing. "But—"

"I know. Our people have studied it, too. The dates, the places, there doesn't seem to be a pattern." She looked up, meeting his gaze. "It doesn't move in what we'd call a logical manner. It pops up, disappears, then pops up half a world away. If the reports are even slightly accurate, the first sighting was in Russia. Then it moved here to Canada. Then to Egypt. Persia. Lots of appearances over oceans. And well, you can see it for yourself."

"I can, but this makes no sense." He'd forgotten to be annoyed with her. Now he was puzzled. Curious. "What are these additional notations?"

"Places where people were missing after the storm. The ones we know about." Sometimes there weren't people to say if people were missing.

His gaze dropped. "So not every time?"

"That we know about." She hesitated, but it had to be asked. "Were you at any of these places? No—I know not when the storm was there, but at any other time, before or after?" Yeah, it was straw grabbing, but straws were all she had right now.

His finger touched the map, tracing a path only he knew. He'd written Dr. Masterson about the weather, but hadn't mentioned where he was, or outlined his route. It was as if his mind had become hyper focused. So much so that when he'd gone airborne, only the weather had mattered. Where he was for that weather was incidental. Only he knew exactly where and when he'd been. His finger tracked up to New York—he'd met Dr. Masterson there and bonded with him—then across the Atlantic. The only thing they knew for sure, he'd avoided this continent, avoided Galveston and the Southern United States. For four years. And had only been coaxed back by The Storm and Dr. Masterson's urging. It had taken some tricky dimensional travel, getting Dr. Masterson here in time to meet Ernest the first time. She still wondered what the good doctor thought of him, though this was his only Ernest.

Finally, his finger stopped moving.

"Here. This is the only place both I and the storm have both been. But my visit was long before the storm, months before."

She leaned over. "Persia." She rested her elbows on the map. She could tell by the way the symbol was rendered that The Storm had not been in Persia in this dimension. Which was, well, puzzling. "Okay, after Persia the storm tracked toward England. But stopped off the coast." This time in this dimension. "A ship was lost at sea that may or may not have been because of it. There were no survivors to talk to. The *Covington*."

He frowned again. "What's the date?" He bent to look for himself, then sat back with a frown. "There was a geological expedition that I spent some time with in Persia. One of them was from England. I think, I could be wrong, but I feel like I heard that ship mentioned. Maybe by him." He looked up. "This is ridiculous."

He was right, but it was the first time there'd been even the faintest sign of anything. "Isn't that how science works? You start with an inkling, a theory, and then follow it until it leads to something else or fades to nothing?"

He made a move that could have been agreement. "This is something less than inkling. Even for science." He frowned deeply. "That you consider it one tells me you have very little."

"I think saying we have very little is being wildly optimistic of you." She grinned at him, and he almost grinned back, before he caught himself. "Look at the other locations. Any of them related to your expedition? Or your travels? What about the people?" She handed him a list. "This is very incomplete, but any of the names ring a bell?"

He was quiet for long enough that she felt hope wane. Then he spoke, slowly, almost absently. "They were from all over, but I don't know precisely where—possibly Canada. Mexico. England... I wasn't overly interested in their names or where they came from as I was in what they were doing. They were geologists, surveyors, drillers—they were looking for petroleum reserves for the Shah."

So it was easy to keep them at a distance if he didn't know their names and stories, keep it all from getting too personal. Personal led to personal questions. Questions that would remind him of what he'd lost. He'd surrounded himself with automations to get away from the personal.

"What prompted you to stop there?" He was going to have get over answering questions, even personal ones.

He shot her a look. “I was in Tehran getting supplies. Someone, an agent perhaps, asked me to check on them. They'd missed a meeting or something. Might be low on supplies. I had room, so I took some things for them, in addition to my own, in case I ran across them.”

Prudence straightened. “Really?” She didn't remember enough to know when they'd started hunting oil in the Middle East. Hadn't needed to know. An Agent had a crap load of stuff to remember, and what they couldn't remember got pre-loaded into their gear, but it wasn't like all of it could be stored there. They had to pick and choose. Her upload had focused on the meteorological, not geology. “Was it a company or individual behind it?”

The question was more rhetorical, so she surprised at his reaction.

He stiffened first, then shifted in his chair. “They weren't saying, it was confidential, to prevent speculation, but one of them slipped and mentioned a name. I gave them my word I'd forget.”

Which would be the one thing to make sure you never forgot.

“Person or company?”

Ernest's lips thinned again.

“We need a name.”

“I gave my word.”

“I promise not to speculate.” She waited a few seconds. “It could be important.”

Still, he didn't speak. Men and their word. Just then Winston appeared in the doorway.

“There's a gentleman to see you, sir.”

Ernest blinked. Prudence was more like stunned. They were on an island, for Pete's sake. How had he come? Airship? Boat? Magic carpet ride?

“To see me? How—” He stopped, obviously realizing the automaton wouldn't have an answer to the question. “Did he give a name?”

Winston held out a calling card. Ernest took it, stiffening. He looked up. "Tell him we'll be with him shortly."

Only after the automaton had left did Ernest hand her the card. Prudence read it.

"Lord Henry?" She started to shrug, then it hit her. "Him? He's the guy?"

Ernest nodded grimly.

Prudence tapped the card against her thumb, her brows pulling together in a frown. How she'd have loved some time with the Agency's main difference engine. "Interesting."

"Interesting? I didn't even know I'd be here. How could he—" His voice rose a bit. "It's impossible—"

"Number two on our impossible things list?" Though not the number two she'd been expecting. His arrival was very interesting. And disturbing. Profoundly disturbing. She might just might, be in more trouble than she realized.

Chapter Eight

LORD HENRY'S threads said he had the bread to splash around, though in this time it wasn't that hard to know what was cool, so Prudence wasn't impressed. What he couldn't buy was the build to be boss. He was a black scarecrow, with his stick arms making awkward lines and angles. Even the sassy cane looked out of whack.

He was a bit too everything for the shabby drawing room. And his arrival had thrown her off balance enough to let the slang creep in. She tried to get her head into 1905, while he did the polite—greater to lowers—bow, first to her and then to Ernest.

The way he looked at Ernest was a bit too cranked—too—well, she didn't know what they called it in this time. Crazy maybe or unnaturally excited. Whatever it was called, it kinda creeped her out. But his lack of distinguishing attention to her, gave her time to eyeball him. Goodness, the guy was downright hideous. Oily, coarse nest on top of that misshapen head. Hair sticking out in all directions. Pocked skin and a nose and mouth in fierce competition for ugliest. With his ears edging into a close third.

He was kind of a caricature of himself, one that had accidentally come to life.

"Dr. Masterson thought you might be here, Dr. Warren." He paused, letting his gaze trail briefly around the space, arching his brows just a bit, as if he wanted to ask why but was too polite to do it. His accent was upper class British. "I am pleased he was correct."

And apparently, the doc had a crystal ball. If Masterson had sent Lord Henry here. Anyone with money could follow the crumbs between Ernest and Dr. Masterson. It was the same calling card she'd used. Copycat. So far she wasn't buying what he was selling. Significant to her, without offering clarity, there'd been no sign of a Lord Henry around the other iterations. Didn't mean there was something wrong with the guy, even if she really wanted to think there was. Cat could be creepy without being creepy.

Ernest stared at him for several seconds and then muttered, "The storm..." He gestured vaguely toward the Gulf. He opened his mouth, probably to ask how either man could have known he'd be here, but closed it, most likely caught in the 1905 manners trap.

Prudence would have liked to know, too. She hadn't told Masterson where she planned to set up shop.

As if he'd become aware of her, Lord Henry looked her direction, and she discovered there was one thing about him that wasn't ugly. His eyes. Though cavernously deep set and dark, there were lights in there that were kind of mesmerizing. They saw what they looked at, which eyes didn't always do, particularly when directed toward women in this time. Or her time, she reminded herself a bit wryly. There was a fathomless quality to them, leaving her the sense that if she looked long enough and hard enough, she might learn something. Maybe something amazing. Find out deep secrets. Become...better. She felt herself leaning and rocked on her heels.

That woke her up. She dropped her lids over her eyes and took a couple of deep breaths, her clasped hands tightening briefly. The dizziness faded. The wood beneath her feet firmed once more. Carefully, cautiously, she raised her lids, though not all the way up.

Lord Henry extended a claw-like hand her direction, giving Ernest a look of inquiry.

Ernest jerked at this nudge to his manners, but Prudence beat him to it, the instinct to hide her true name well honed by her many dimensional transits. That she'd told this Ernest her real name was another protocol breach, but she hadn't thought it would matter. "Prunella Smith."

Lord Henry had his eyes on Prudence, so he missed Ernest's jerk. Prudence stifled a sigh. He was brilliant, but not subtle.

Prudence let the clammy claw close round her hand. At least his slight bow gave her break from the eyeballs. The small shift of air brought her the scent of him. The first wave was pleasant, but the follow-on, not so much. It wasn't bad enough to repel, but there was something about it that made the hairs on her arms lift. She didn't like him, so it wasn't easy to decide if that made him dangerous or just one of those guys women didn't like, well, because. He made a great alternative to Ernest as the bad guy.

No wonder they didn't want agents to get emotionally involved.

"Please," she gestured toward a wooden chair next to a small couch. "Won't you be seated?" She clasped her hands back together and added, hoping it didn't sound too afterthought-ish, "Can I get you some refreshment?" Jeez, she hoped he said no. Though maybe one of the automatons would know how to manage that?

Lord Henry sank gracefully onto the wooden seat, one hand still clasping the cane, but shook his head. "No, I thank you."

Something in the way he glanced around seemed to hint that he couldn't imagine good refreshment in this place. He had a point. Even the water was on the brackish side.

With a look at her, Ernest went to the bench. She followed him, and they sank down together.

The silence grew slowly awkward.

Prudence liked awkward silences. There were things to be learned in them if one didn't get sucked into awkward. Ernest didn't love it. She knew that without looking at him. Lord Henry didn't like it but didn't seem to fear it. She could almost see wheels turning inside those eyes, his brain searching for a way past the current impasse without him being the one to give ground.

Lord Henry finally cleared his throat, as if acknowledging someone had to do speak. Shifting slightly so that he faced them more fully, he said, "I am glad you made it safely ashore, Doctor."

"Thank you." As if even Ernest realized this was not enough, he added, "I was rather concerned." Another pause. "And relieved. To make it, I mean."

Not a great example of clever, though ultimately just as unrevealing. Prudence had to lower her lashes quickly and keep a firm grip on her lips to hold back the grin.

"I understand you spent some time with my expedition while you were in Egypt?"

"They were...I found them very...they were most kind."

A very slight smile curved the edges of Lord Henry's thin mouth. "I believe you were the one who was kind. They were in need of resupply. I am grateful for the assistance you provided." A pause. "And I am grateful you did not mention my involvement. Your discretion is most appreciated. When it is known that one is willing to fund expeditions..." He made a gesture with his hands that seemed to say he'd like to fund them all but couldn't.

"I imagine...they didn't..."

Lord Henry held up a hand. "Cleveland was worried, but I assured him that I did not mind you knowing. In fact, I am most interested in your researches, Doctor. I have my ticket for your presentation at the Institute."

He moved the cane to a position between his legs, resting both hands on the head. Between his hands, the silver gleamed, something that could have been beak or possibly a muzzle stuck out of the gap created by his palms. Did she see jeweled eyes? Okay, her imagination might be running a bit wild.

Prudence felt the slight withdrawal from Ernest at the mention of the Institute. Lord Henry saw it, too.

"You are going to speak, are you not?"

With an effort, Ernest nodded. "Of course. I just wonder how useful my research is..." His voice trailed off, and his focus shifted as the scientist came forward. He looked toward the Gulf, toward the incoming storm.

Lord Henry's eyes gleamed. Was it her overactive imagination that the cane head's eyes did, too?

"So the storm you ran into was The Storm?"

Ernest still resisted. It showed in the tightening around his mouth, but he admitted, "It did exhibit some unusual features."

Lord Henry leaned slightly forward, the knuckles of his hands turning white. "I've read about it, but have not yet experienced this storm."

Stick around, Prudence wanted to tell him, even though she didn't want him to stick around.

"You may have the opportunity," Ernest said, "if it persists on its current course. Though, from what I've read, it can dissipate quite suddenly and without warning."

Lord Henry leaned back again, with an effort at calm or so it seemed to her. "Yes, I have heard this as well." He too looked in the direction of the Gulf. His gaze returned to them, tracking between them as if trying to figure out their relationship.

He looked at Ernest. "I had come to offer Dr. Warren a share of my modest hospitality..."

There was a hint of a question in the comment, but also some irony, Prudence noted, fighting a grin once more. It would be hard to get more modest than this.

"I am, it is most kind, but..." Ernest shifted on the wooden bench.

"It's very hard to separate Dr. Warren from his airship," Prudence said, to help him out and to add an element of respectability to the situation. Not as much as Ernest would have liked, but as much as she could give him. "Though I believe he was glad to have a break from sea rations."

"Ah," Lord Henry relaxed some, which was interesting. Why did he care?

The air in the room was thick with nuances and tension and questions. All kinds of crap. So much it was hard to sort through. There was menace in there, too, but she wasn't sure it came from Lord Henry or the storm tracking so steadily toward them.

"When do you think The Storm will arrive? If it does," Lord Henry amended.

Ernest hesitated, a frown creasing his brows. "Possibly by morning."

Lord Henry rose, prompting them to rise as well. "Then, with your permission, Miss Smith, I would like to return in the morning? I can imagine no greater pleasure than experiencing it with a scientist of Dr. Warren's caliber."

Prudence did not want to be here to greet the storm, didn't want to face it with those eyes watching her, but what could she say? She glanced at Ernest, then managed a polite smile. "Of course, sir. We'd be pleased to have you join us."

Chapter Nine

Even after Lord Henry's departure, Prudence didn't speak. Ernest liked how quietly she sat. It was unusual in a woman. The crease between her brows seemed to indicate that she was puzzled. He agreed that there was much about the meeting to puzzle, even without the strange storm heading toward them.

"He was," Ernest finally broke the silence, "most unusual."

"Yes."

Ernest waited for her to say more and when she didn't, he shifted to look at her, only just realizing how close they sat on the uncomfortable couch. Her scent, something light and as unexpected as she was, came to him, for once the reminder that she was a woman not causing him pain. He didn't try to discern why. The situation had pulled him out of the past, out of pain, out of everything familiar. Into the unknown. Into the impossible. He found he didn't mind as much as he should.

"You didn't trust him?" The question surprised him. He had not realized he felt distrust until he asked it.

"No."

She looked at him, her face close enough that he could see all the colors in her eyes. And the curve of her lips...

"Did you?"

The question pursed her lips just right for—

He looked away. He shook his head. "No." He frowned. "I'm not sure why." He looked at her once more.

"He's creepy," she pointed out. She sighed. "That doesn't make him the bad guy."

"Bad guy?" He shook his head.

"The villain of the piece."

"Oh." Ernest looked at the door where Lord Henry had left. "He would be excellent on the stage." He frowned as something in his gut twitched. A sense that he'd said something that mattered.

"Do you think he was acting?"

The direct question startled him. "I'm not good—"

"Do you think he was acting?" she persisted, leaning in so that he couldn't look away.

He frowned. "I'm not sure that was—no—not acting so much as he doesn't seem...authentic." He rubbed his aching head with frustration. "That's not right—he was too—"

"Was he," she hesitated long enough he had to look at her, "too real?"

He considered this, then nodded slowly. "Yes." He studied this notion and felt the correctness of it. "How strange. Can one be too real?" One was either real, or one was not. Or so he'd always believed. He could concede that his notion of reality had been tested in the last few hours.

She tipped her head to one side and seemed to seriously consider the question. Then she made a face. "Before today, I'd have said no, but today we're working on believing impossible things."

Impossible things. It was the first indication that she experienced difficulty with them, too. How many had he been forced to deal with so far? He rubbed his forehead. "It is stuffy in here." He was used to fresher breezes, airship life. This place felt closed in, confining. His thinking processes felt as clouded and confined. "I need to walk." To pace is what he meant. How many miles had he logged pacing the perimeter of his airship?

"Let's go for a walk. With the beacon lit, it's clear enough out there, though we'd better stick to the boardwalks." She sighed. "Pacing sometimes helps with thinking."

She appeared indifferent to the proprieties. Technically they were not alone in this cottage since the automatons were close by. Her confidence in his integrity both pleased and puzzled. It implied, once again, equality between them. Not assumed either. Expected was closer to what he sensed from her.

She didn't speak again, not until they strolling along the boardwalk toward the lighthouse. The tide was still out. In the diffuse light from overhead, he could see the high water lines on the rocks. He dimly perceived the hull of his airship, almost motionless in the unusually still air. The pier jutted out from the other boardwalk, the one from the lighthouse. It looked to be sturdy, as was his ship, but if the storm turned severe...he wished he could move it. He frowned. He could have his ship handlers take it into Mobile, but if the storm moved inland, they did not have the programming to react suitably to the threat.

"You're concerned about *The Weatherman*."

Did she read minds? He nodded.

"I am concerned about all of us," she admitted, with a sigh. "I wish we had time for a tour. It looks to be a fine ship."

He opened his mouth to offer but closed it again. In the dark —no, that was taking proprieties too far. He realized she was the first woman he didn't mind boarding his airship. Because he didn't feel guilt, he looked up, anxious to change the subject.

The glow from the lighthouse beacon was different from moon or starlight. The light was softer. Brighter, then softer as the Fresnel lens turned in its unique pattern so that ships would know which lighthouse this was. He studied it, interested to note that it created a sort of umbrella of light overhead, the effect quite interesting. He'd only ever seen a beacon from out over the ocean, he realized. The change in perspective was intriguing. That thought made his instincts twitch. Or flinch. His perspective of many things was currently being challenged. A change of perspective was also the reason he'd left, or so he'd told himself. How much had it actually changed, he wondered now.

"I've become quite fascinated with lighthouses," she murmured, her gaze also directed upwards. "They are beautiful and...hopeful."

Hope. He'd been hopeful once. Felt it stirring inside now. In this light, with her chin raised like that, she was quite beautiful. For the first time in four years, he felt no guilt at admiring another woman. She was so different it was impossible to compare, he told himself. She turned, her chin lowering as her gaze met his, and he had the most extraordinary feeling that he'd been here—in this moment—with her, before...

He swayed as the horizon tilted, and she grabbed his arm to steady him.

"Are you all right, Er—doctor?"

He covered her hand with his, and with the horizon in a slow spin around them, said, "You called me Ernest....before."

The pause was long, but she didn't look away or appear startled or any of the things she should have. Instead, she looked intrigued. "Did I?"

"But—" he rubbed his head. "How can I have this memory when I know this is the first time we have met? Or met here...I have certainly never been here before."

She steered them toward a spot where a mass of granite offered seating. He sank down, needing the respite for knees that did not feel like his own.

"Your...memory is specific to this place?" she asked.

He examined it and finally nodded. "Yes, though it makes my head hurt thinking of it." He looked at her. "How can that be?"

"That would be another impossible thing," she said ruefully. "I call them the impossible possible since they clearly happened."

"You don't remember?"

She shifted a bit, her mouth pursing for almost a minute. "I don't remember, no, not yet, anyway. Time is, well, fluid. That makes us uncomfortable, so we call it *deja vu* or past lives or—" she grinned, "a knock on the head. It's only if we open ourselves up to the impossible that we begin to accept...more and be surprised less."

He stared at her, not wanting to believe—his mind whirled, but his heart said he knew this woman. Against logic and science. Knew her and—as if in a dream, he lifted his hand and touched the side of her face. She didn't flinch or move back. Just watched him, with curiosity in her gaze, but something more. He looked deep and found cautious hope in the brown depths. Hope. It was quite...contagious tonight. He thought he'd discovered all the colors in her eyes, but he found more in the beacon light. Golds. Greens. Dark and light. So much more light than dark.

In some strange fashion, he felt she *saw* him, all the parts of who he was, even those he was less enamored of, and still liked him despite...

His hand slid along her cheek, then around to the back of her neck. Her skin was softer than the finest silk, the stroke of her hair to the back of his hand sending light into his long darkened, hopeless heart. *Hope.*

He didn't think. Instead, he bent toward her, noting that her head tipped to accommodate his approach, angling just enough so that they came together, it felt not new, but as if it had always been like this between them. It felt more natural than he'd ever felt with Ophelia...

The rightness of holding her, of kissing her shocked him. He pulled her closer, diving deeper into the kiss—he demanded, and she did not seem to mind—

There was the sound of wood against wood as if something

had hit the pier. He jerked back, waited for the guilt, but it stayed at bay. His hands lingered on her shoulders. She stared at him, her chest rising and falling rapidly, her eyes big enough to fall into again.

A wave hit his legs, drenching him to his knees. She looked down, so he did, too.

"The tide is coming in," he said.

She frowned, shook her head. "It's not time for the tide. And it shouldn't reach us up here in any case."

"Then what?"

"Could be storm surge. Must be."

Storm surge? "Do you refer to storm tide?"

Before she could answer, they heard footsteps thumping on wood. The sound changed some when it moved to the boardwalk. Now they felt the vibration of the steps on the wood under their feet.

"Sir?"

Ernest huffed out a sigh and let his hands fall away from her. "That's Niles. One of my ship handlers."

"Niles." Her lips curved a bit as if something about the name amused her.

The boardwalk shuddered. Like it had been waiting for his attention, the wind freshened. Only instead of the scent of the sea, it carried the brimstone smell and taste he recalled from the storm.

She rose to her feet and turned to face the Gulf, her chin lifting toward the dark horizon. "I think your storm is almost here."

"That's not " He stopped. He would have to come up with a new definition of "possible," it seemed.

The automaton came into the light cast by the lighthouse beacon.

"One moment all was quiet and then," he blinked. "And then it wasn't, sir."

"Crispin?" Ernest asked.

"He was below deck."

"Is *The Weatherman* all right?"

"It was intact when I left it, but..." the automaton stopped, lacking the programming to explain what had happened, Ernest surmised.

"The water is rising fast," Prudence said. She glanced back toward the cottage.

The glow from the cottage windows somehow appeared further away than logic said they could be.

Ernest studied the horizon again. His impulse to send Niles back in for his counterpart died. Outside the beacon-cast light all was deeply, ominously dark. "It almost appears that—" he stopped, unable to say the words.

"Like the beacon is keeping it back," she finished for him. "That wasn't on my impossible list." She sighed. "If it is holding it back, we'd better make sure it stays lit until—"

He looked at her. "Until?"

"We come up with a plan."

It was, he realized, possible to admire a person's optimism and be astonished by their apparent insanity. One did not plan for this. One fled storms. If one could not, then one rode them out hoping, praying to survive. And then one dealt with what came after. The guilt, the pain tried to come back, but there was not the time for either. That would also have to wait for after...

Chapter Ten

"If the water keeps rising, the cottage will not be high enough," Ernest said. He looked around as if he expected another dwelling to appear.

Prudence agreed with the unspoken wish, but yeah, this island was going, going, almost gone, even before this crazy storm. This storm surge, if that's what it was, wouldn't help with the erosion problem.

He glanced back in the direction of his airship, but said, "We should seek refuge in the lighthouse."

"Yes," she agreed, her mind still kind of stuck on how he could feel like he knew her. She hadn't seen that coming. And how was it possible that he remembered her calling him Ernest? He'd never allowed that liberty in any of the other dimensions. And he believed it was place-specific. That seemed to indicate some kind of time shift, something she'd read about in some of the more obscure texts in the Agency library. But that didn't explain how he could remember something that hadn't happened yet.

She mentally backtracked, trying to recall the right research branch. Been easier with the difference engine, but oh well. Okay, what if Ernest wasn't *the* anomaly? What if the whole dimension was a...closed system? She thought that was the term she'd heard used. Masterson had written the tract, at least she thought it was one of his. The little she remembered, he believed there were dimensions that didn't spawn other dimensions. It was all theory because no one had traveled into one. Or gotten out if they had? Why did her thoughts keep wanting to circle back to her MIA grandfather? It was unusual for their kind to disappear without a trace. Odds were, even if an Agent got lost in a dimension, someone would stumble on signs of them eventually.

But if she'd met Ernest before, then that seemed to indicate, her mind hunted for a word that would fit, a different type of anomaly. What if it wasn't so much about time travel, but a time... repeat? With some memory bleed through? But why would time replay? She didn't even ponder the how because she was already way past the known unknown.

Okay, say it was repeating, or replaying? Replaying until—what? Well, The Storm was seriously messed up. Or something was. What if this were some kind of correction loop? Early Pinkertons certainly believed in a Higher Talent. Some skepticism had crept in through the years, but the principle of doing no harm was still the standard of conduct within the Agency. She'd heard rumors of rogue Talents, but stuff like that was not on her reading list. No one said so, but it had to be the reason Agents traveled with weapons.

This Storm was doing damage across dimensions. If there was a Higher Talent...or was time was trying to self-correct? Or man-made tampering gone wrong? Did it matter what had caused this?

It was all pretty far-fetched, but people in the Agency had postulated stranger things on less data.

"I wish I had my gear," he frowned, "though I'm not sure it would be of any use. It was not functioning normally before I went under the storm."

"I had your crew move your stuff ashore while you were unconscious. It's in the lighthouse." She said this somewhat absently, lifting her sodden skirts as a wave crashed just behind them, the water splashing in a high arc as it hit rock. "We need to burn rubber, or we're gonna get creamed."

She got a weird look and realized she'd lapsed into her time talk. She grimaced. "We should hurry." She gathered her skirts into an untidy bunch and legged it. It wasn't as bad as it could have been. She had her working uniform under the 1905 duds. Ernest kept up, half lifting her up the porch stairs as water tugged and boiled across the boardwalk. Niles managed to keep up with them, which was a bit surprising. Running was a fairly complex activity.

Inside the cottage, Ernest called to his other two automatons.

"I'll be right back," she told him. She dashed, well, dragged herself and her sodden skirts, into her bedroom, tugging at the various things keeping the now amazingly heavy skirts attached to her person. Luckily the snaps and such weren't of this time, so they were easy to disconnect. She let the mess drop with a thump, stepped out without even trying to kick them aside. She'd probably break something. She dug under the bed for the weapons she'd hidden there and quickly strapped them on. The rest of her gear was easy to stow in the bags attached to her utility belt. Ernest was about to get shocked round—what round was this? She'd lost count—when he saw the pants, but she was not going to battle with a storm in a dress.

She took time to check on Octavius. He seemed to be all right. Looked like he'd gone deep. Their connection wasn't solid. That worried her. If this was an anomalous dimension, and she lost her familiar, she and gramps would have more than the Talent in common. Her father had been worried when she bonded with Octavius. It was very unusual to have a sea-bound familiar. In fact, the only other person to have one was that MIA grandfather. Not a good time to think about that.

Back in the parlor, Ernest waited by the door with his automatons. His eyes widened, but the sudden howl of the wind made the structure shake. He didn't have time for shock or questions. He gestured toward the door, then seemed to realize this was not the time for ladies first. Not that she wouldn't have been okay. She was likely more physically fit than he was. She studied his rear aspect. All wet like that she could see what he had. Okay, after his travel around the world, make that almost equal in the toughness stakes.

Prudence took a position to one side of the door and nodded. He yanked it open, and the wind rushed in like it had a mind and a purpose. Ernest stepped into the gap, and it threw him back, almost knocking him off his feet. His two bodyguards moved to help, kept him on his feet. He looked at her and gestured toward them. She nodded. The automatons should be able to make headway, even against the high wind. Or they'd all tumble together. It was pretty much a do or die moment.

On his order, they moved into the doorway and waited, their aspect a bit bent against the force of the wind. Prudence got behind one, while Ernest positioned himself behind the other. Ernest grabbed her arm and the belt of his automaton. Prudence did the same with hers. Niles came behind them, a hand on each of their shoulders. The automatons linked arms and pressed forward.

It was crazy bad outside and, out on the porch, the automatons ability to block the wind diminished as it grabbed at them from first one direction and the other, though it did help to have Niles steadying them from behind. In the canopy of light from the beacon, the waves tumbled over themselves in their hurry to block their path. The wood of the boardwalk was slick and treacherous as waves crashed over it. Good thing it was a straight shot since it made it hard to see the boardwalk. Out on the water, visible because of the beacon, the darkness looked like it beat against the light. She'd seen some weird crap in her various transits, but this took the prize for sure. She heard a new sound. A sort of weird roar and flashes of orange light flickered against the beacon canopy. Lightning? No. Something was up there. Its shadow moving—no, had to be flying over the canopy. Didn't have the right shape for an airship, unless it was a crazy weird one. It was the right incentive to get her moving faster. She didn't want to be out here if whatever it was found a way in.

Linked together they made their way through the fast rising water to the lighthouse. Weird that the steps weren't under water, but also good. The climb seemed high, and inside, the climb would be higher. This was one tall lighthouse. The wind sound rose in a shriek when one of the automatons managed to open the door. They almost fell inside, then it took all three automatons and Ernest to close it.

The sudden silence was not as comforting as it should have been. Both she and Ernest were panting from the effort. She looked up. The stairs wound up and up the side of the lighthouse, visible from light spilling down from the beacon. She knew from personal experience that there were a lot them.

"While I would like to wait and regain my breath, I am not sure it is wise," Ernest gasped out.

In the dim light filtering down, Prudence saw something coming under the door. She crouched down and cautiously touched it. It felt like water, but...not. It burned her finger tip. Wisps like steam or fog curled against the frame, as if seeking a breach. She stood hastily and retreated.

"Yeah, we should go up."

"Up," Ernest said to the automatons. They responded with an almost human alacrity. He waited for Prudence, then started up himself. The fog curled around the base of the stairs but did not follow them up more than a few steps.

Prudence tried not to count, though her brain seemed to want to. She was not too ashamed to use the railing to help pull herself up since Ernest was on her heels. The landings were the illusion of respite since more stairs waited on the other side. The counting helped a little when she made it halfway. When they reached the watch room, Prudence bent, resting her hands on her knees. Gravity hated her in every dimension, she decided.

Close to the light, it was brighter, but not unbearable.

Ernest leaned against the wall, trying to catch his breath, too, though he did try not to look wasted. It was kind of cute. Prudence didn't have the energy to pretend. She straightened, with an inner groan, and pulled up her googles. She positioned them over her eyes, turned on the glare reducers, and activated the scanning program. Through the sudden stream of data in her peripheral vision, she saw Ernest staring. She extracted her spare pair, adjusted the filtering, then handed them to Ernest. A bit dubiously, he put them on, adjusting the fit.

The automatons had placed Ernest's gear to one side of this platform, stacking it neatly. It was hotter up here and the stink from the fuel source was...challenging. She went to the last flight of stairs and looked up. The glass windows up there weren't even rattling. She couldn't hear the wind howling anymore. In fact, it was so quiet, it was a bit...freaky. There was a kind of boom or slam. Like the door closing. A pause, then the hollow thump of footsteps on the stairs.

She extracted the Raijin-ator 5, which she'd stowed in rig strapped to her back, and took a peek over the edge. At first, she couldn't see anything. Just heard the steady sound of approaching footsteps. Found herself counting again. Ernest came up next to her, giving her long gun a bemused look. He leaned over next to her. The steps kept coming.

"I think it's Crispin." He straightened and looked at her. Prudence arched a brow. "My other ship handler. The one on the ship."

Prudence didn't find that as comforting as she'd have liked. She kept Raijin-ator 5 directed down.

"It's—"

"Crispin, I know." She looked at him. "How did he get off the ship?"

Ernest was quiet, clearly processing this. "You can't just shoot him."

"I wasn't planning to. Order him to stop." If he didn't obey, then she'd shoot him.

Ernest nodded, leaned over and called, "Crispin, stop at the next landing. Go into maintenance mode and check systems." He leaned back.

One more footstep, then silence. She looked at Niles. "Maybe you should have him give himself a checkup, too. I think I'm having trust issues."

Ernest didn't look happy, but he nodded.

"Have him set up here. He can block the stairway." Or Crispin could have his back. "Maybe have your other automatons set up your stuff over there?" The mechanism that rotated the light ran down the center of the lighthouse. Some weights and such. At least they had the high ground. Overhead there was a raucous cry, unlike anything she'd ever heard. Maybe they had the high ground.

"My equipment was malfunctioning as we approached the storm," Ernest said. "I don't know why you bothered to move it up here."

"Was it malfunctioning or giving you readings you didn't like?" She didn't look at him. Her attention was still focused up the last stairs.

"But—" he stopped.

She looked at him then, saw thoughtfulness creeping into his gaze.

"Stop looking for the expected and see what's actually there." It was like the first rule of dimensional travel, but he wouldn't know that.

"What are you going to do?"

She put a foot on the first stair. Dang, that was a lot of heat coming down the well. "Going to take a gander at what's out there."

He opened his mouth, either to ask what or protest.

"I'll be fine." Maybe. Maybe not. It had to be done. She pulled the Paraton-ator 20 out of the holster and handed it to him. "This is the safety, that's the trigger." He looked pleased to have a weapon. She just hoped he knew who to point it at. She grinned at him and was startled and delighted when he grinned back. It was such a spontaneous and natural expression. She'd never seen any Ernest do that before.

Then it was gone. His gaze was serious. "Be careful out there."

She nodded, even though she was obviously never careful. If she were careful, she'd be back home, working as a teacher or nurse or married to a guy with a normal, 1950s kind of job. So she grinned again and headed up, with the long barrel of her Raijin-ator 5 pointed down, her other hand gripping the hand rail. The heat and smell built with each step up. Of course, it would be hot. The sun beat in all day, and now the lanterns with their stinky fuel got to weigh in. There were vents, but yeah, still a mostly closed space. She circled the huge light to the balcony access and eased it open.

Nothing. Not even a wisp of moving air.

She was definitely getting the impression this was The Storm, not *a* storm.

She eased partway out, blocking the door's closing with her body. Just in case the storm changed its mind. She fine tuned her goggles, adjusting them to this light, and studied the scanning data. Still no signs of dimensional travel, though she had to keep a tight rein on her instinct to flee. She scanned the horizon, taking it slow to give her equipment as much raw data as possible. If she didn't make it and someone came after—assuming someone could—they could download the data from the drone.

A shadow passed overhead, and she looked up, blinking as it circled the tower, then appeared to settle on the roof next to the lightning rod, its wings flapping once, before being pulled in against its body. It looked to be the size of a largish dog. Something that looked like claws curled over the edge of the roof, not far from where she stood. As if it sensed her, it angled its muzzle her direction, its eyes glowed red and round, and it blew a golden-red stream against the canopy. A golden-red stream of...fire?

She sensed movement and turned to find Ernest standing behind her, looking up with his jaw a bit dropped. "What in Hades is that thing?"

"I think it's a dragon."

A DRAGON?

"You can't be serious," Ernest said. She'd asked him to believe six impossible things, but a dragon? That was one impossible too far.

"At least there is only one," she murmured. "And it's not that big—"

There was raucous cry, further away, and the dragon lifted its head to answer it with one of its own.

"Oh man, shut my mouth. This day just gets more and more... interesting."

She flashed him a grin, and he wondered if part of interesting was the kiss they'd exchanged. Who was this woman? She wore pants. Carried weapons like nothing he'd ever seen or heard of. Spoke of the impossible and dragons and storm stopping. And she kissed like an angel. She made him feel alive. It would be most unfortunate if he died, just when he finally wanted to live again.

He sounded his heart for guilt and found none. There was not time for guilt or kisses. Even without the dragons, there was the strange storm bearing down on them. This structure, which had seemed so sturdy from down on the beach, now seemed as precarious as his airship. Despite that, he felt exhilarated, happy to be here with her. The danger sharpened all his senses. Made him feel more alive than he'd ever felt before. He didn't prod the old wound, but distantly knew he'd never have been able to share a moment like this with Ophelia. It did not matter. She had no place here. Ophelia belonged to another life, a different time. It was only now, staring into death's maw, that he realized how very little he'd lived the last four years. Or the years before that. All the places he'd been, the things he'd seen, how sad to realize now that he'd done them through a filter that kept him from feeling. One had to live, to learn. To *live*. Not merely exist.

"Look at that."

It was not easy to do, but he dragged his attention away from the dragon. She pointed down, so with some concern he looked down. And blinked. The water had retreated, exposing the granite chunks of rock once more. The roiling surf had also calmed to the point where it was placidly lapping around the pilings of the pier. His airship was still outside the range of the beacon, so he could not tell if it was intact or not. The wind had died down. A shadow passed over, tracking across rock and water in a slow circle, the wing span about enough to keep a large dog—or dog-sized dragon—airborne.

"The equipment has been set up, sir." Fred spoke behind him.

Ernest looked at Prudence. She shrugged.

"Let's go have a look."

To his relief, she climbed back inside, making no demur when he closed the door. They followed Fred back down to the watch room, where he saw that they'd arranged his instruments on the floor in a semi-circle to one side. There were additional crates stacked to one side. He hoped those contained his research notebooks and books.

She crouched down, studying each one in turn. He had that flickering sense of having done this before with her. Like two images superimposed, but one faint and somewhat out of focus.

"They are misbehaving, aren't they?"

They were indeed. His sunshine recorder believed there was sun to record. His hydroscope could not decide if there was or was not moisture in the air. His hydrodeik showed similar issues with the humidity. And his barograph said the pressure was rising and falling, then rising again. His barometer showed dropping pressure, but how was he to know if that was a correct reading?

"What's that one? It's not doing anything."

"That is my Blizablighter 3000," he told her. "It is not activated."

"Really?" She shifted position so she could study it, touching it tentatively with the tip of her index finger. "Bliza-what?"

"Blizablighter 3000."

She turned and looked at him for a long moment, seemed about to speak, but didn't. Redirected her gaze back to the Blizablighter 3000.

"What is wrong?" he asked.

"Name just seems a bit odd for you."

He frowned. Now that he considered the name, it was an unusual choice for him. He felt that flicker again as if seeing something that could not be, a sense that he had not been the one to name it. Only he had been alone with his automatons when he had built it, then named it, so of course, the name came from him. Perhaps it had been an unusual day?

"What does it do?" she asked her words almost an exact echo of those in the flickering memory.

"It acts somewhat similar to a lightning rod. But instead of directing the electricity into the ground, it collects it, stores it for later use."

"That thing stores electricity?"

"Well, I've only tested it on small strikes and only after testing it with small amounts of artificially generated electricity, but yes, so far that is what it does."

Outside, the dragons roared, first one, quickly followed by the other. Prudence looked up, then at Ernest. "Are these the only instruments behaving badly?"

Ernest pulled out his pocket watch and showed it to her.

"Did it start when you went into the storm?"

He shook his head. "Before."

She looked interested, taking the watch in her hand. She extracted her watch and held it up. It kept time with his.

"That might explain how the storm got here so quickly," she murmured.

He frowned. "I do not understand your meaning." Or her reasoning, if that is what it was.

"You're still assuming this stuff is all wrong. What if it's right?"

"Right...how can that be right?"

"Well, what if time is moving faster? Wouldn't that make all the instruments react differently? It would also explain how you and Octavius got here so fast."

There was no sign she jested. He opened his mouth to protest such an egregious lapse of logic, but heard instead, a distant echo of his voice doing just that. He clamped his lips together as the words were followed by the sensation that he'd been wrong then. Dangerously wrong. That he could not afford to be wrong again. It was so important he not be wrong...

"I suppose the storm could be causing it. Or something in the storm. Or the storm is taking advantage of time moving faster," she murmured, apparently oblivious to his failure to comment. Her gaze moved slowly from one instrument to the other, then stopped when she came to the Blizablighter 3000. "What's your battery? Your storage device for that?"

The feeling surged once more, the effect almost like that of glass shattering, but was quickly gone. Too quickly. He needed to know what it was he needed to know—

He shook his head and knelt next to her. Just this simple action smoothed or soothed, that feeling. It felt oddly new. He touched the mass of material he'd suspended beneath the collecting rod. It was multi-colored, but he'd noted that it became more green and gold when holding more charge. "This mineral both conducts and contains the energy."

"I've never seen anything quite like it," she murmured, touching it as well, her hand brushing against his for a brief moment. Her eyes widened. "It's warm. And a bit...tingly."

"That is from the charge I collected in Europe. It dissipates very slowly. It is not safe to touch it right after it receives a strike, even a small one."

"Where did you get it?"

"In...Persia." His voice and thoughts slowed, almost stalled. "One of the drillers gave it to me when he saw it interested me. I believe it might be a meteorite. They found several at one of the sites they were studying." He'd had a sense then that he should have handed it over to one of the geologists, but also had a sense he should hide it. That was the one he'd acted on, though he did not know why. Minerals were outside his field of research. But he did know enough to know it was probably not of this earth. To salve his conscience, he'd tested it, discovered its conductivity and ability to hold the charge for so long. He'd been looking for something that would attract and defuse the power of lightning strikes, perhaps harness them into something useful. He'd planned to show this invention to Masterson.

"Persia." With the tip of a finger, she traced the wires he'd attached to direct the strike into the mineral. "How much electricity do you think it can store?"

"I have been unable to determine the upper limit," Ernest admitted. "I had hoped to sail under a larger storm during the crossing." He hesitated, then admitted, "There was considerable lightning activity in the storm. I was...tempted."

"I imagine you were. No sign of lightning out there now" She glanced over her shoulder in the direction of the storm. "Interesting."

"Why interesting?"

She half shrugged. "It seems able to...respond...think...plan. Maybe it sensed that the lightning would be a temptation for you? It did seem rather interested in you, you'll recall."

Though he still felt incredulous, it was getting easier to navigate the logic shoals. He was not sure if that was good or ill. He could admit to himself that the way her mind worked was curiously fascinating.

"You believe it was," he had trouble saying the words out loud, "a trap? That does not seem—" He stopped because using the word possible had lost clarity for him out here in the shoals of this new reality.

Prudence shrugged. "Or a test. It is the only thing here that no other meteorologist would have, is it not?"

He could not dispute this. He had drawings and ideas but had not felt like he had anything new to offer until the Blizablighter 3000. And truthfully, he was not sure how that helped the study of the weather. Behind that was a truth he had tried to stay ahead of. That perhaps he was not meant to be a weather scientist.

"I am not the weatherman you hoped for," he said, heavily, rising to his feet. He pushed his hands through his hair and down to massage the back of his neck.

She stood up, too. She was tall, but not too tall. Her figure pleasing and sleek in her men's attire. With her back to the center pillar, she lifted her goggles and studied him. "If whatever it is that's out there wants that mineral, then it's possible it's a," she paused, then finished, "a wolf in storm's clothing."

"What is it?" he burst out, yanking the goggles off his face, too.

"No clue," she admitted, "but we need to make sure it doesn't get it. Because what it's doing without it, it's very bad. Not just for you, but for everyone."

He looked at her and asked because he had to know, because he'd asked it before, "Who are you?" He wanted to stop there, but couldn't. "What are you?"

Chapter Eleven

PRUDENCE GLANCED AT HIS AUTOMATONS. Wasn't sure she wanted to have this discussion around a couple of machines who couldn't forget what they heard. "Let's check the lamps," she suggested, gesturing to the stairs. It was not her first choice of a place for a chat, based solely on heat and stench. All that glass, too. But if they went down, they would run into his other machine.

He hesitated, then nodded. She felt oddly conscious of him following her up to the lamp room. Its Fresnel lamp was a thing of beauty, the numerous facets of its glass reflecting and beaming the light out into the night for the ships that might be passing. An engineering marvel, even in her time. The light in the night in this one.

She sighed and put it at her back. Her fingers danced on the buttons of her right arm instrument panel, activating the sound dampening field. It might be detected by whatever was out there, but that seemed a better option than getting overheard.

Ernest grabbed her arm. "What did you do?" He touched it, but of course, it did nothing.

"It creates a sound dampening field so that no one can overhear our conversation."

The scientist warred with man in his expression. "How is that possible?"

She studied him, wondering how much truth he could handle. "Do you have that feeling that we've talked like this before?" she asked, stalling for time, for guidance.

He shook his head. "I did, down there, but it went away."

Okay, that was interesting. "When did it go away?"

He hesitated, then met her gaze with a directness that startled her. "When I started to believe you."

She felt her jaw sag and snapped it up. "Oh. Well, that's... good." Amazing was what she almost said. He'd come further than she'd thought possible. And in a shorter time. The key was the *deja vu,* she decided. If they were in some kind of dimensional time correction, then maybe they'd changed something. But they could still mess up. What was the right move? Right moves? And what would happen if they fixed things? Would this Ernest cease to exist? That was an awful thought. She liked this Ernest better than the others. Liked the kiss. Would she forget again? That was sad, too.

"What's wrong?" he asked.

Everything, she wanted to say. She didn't. "I'm trying to figure out how to explain..."

"The impossible?" He smiled a bit wryly as he said it.

"The wildly impossible."

He appeared to think about this, then said slowly, "I think I can take the wildly impossible as long as it is the truth."

She looked at him. He'd know now. She didn't know why or how, but he'd know if she lied to him. "Okay. The truth then." Chances were, neither of them was going to survive anyway. And if they didn't, would they have to do this again? How many times had they done it?

"The truth then." His gaze had a new steadiness, a new way of seeing her. "Who are you?"

She met his gaze, took a breath and said, "I'm Prudence Pinkerton, an inter-dimensional scientist, sent here to stop this storm from eating all of time and everyone in time. For all time." They weren't sure about that, but that's where it seemed to be heading.

He blinked once. Then again. And yet again. The silence grew, but oddly enough Prudence didn't feel awkward. It was a lot to take in. She respected the time he took considering it all. He took a deep breath and blinked once more.

"Inter-dimensional?"

"There will be this scientist in a few years who will postulate the theory that we exist in a multitude of dimensions, each one playing out a multitude of choices. A person goes left, instead of right and a new dimension forms because they made a different choice. Met someone. Didn't meet someone. Instead of the change rippling forward through that time, a new timeline spins off, leaving the—for want of a better word—the original timeline intact and unchanged. For most people, it will remain a theory." She paused, but he didn't speak. He didn't even blink. "For me, for my family? It's not a theory, it's our reality. We, most of us, have the ability to travel through dimensions and through time. We don't know why or how this happened. We're not even sure when it first manifested. All we know is that it doesn't activate until puberty."

He looked...unconvinced.

She made a face. "My family used to call it 'the gift.' When we realized it wasn't always one, someone changed it to the Talent." She shrugged. "If you think about it, one person is born with a talent for music, another for art, and someone else for inventing. No one thinks that's odd." She flashed him a smile. "It's not very comfortable having any talent, is it?"

That hit home, and he nodded. "No, it's not." He looked around. "So this is just another dimension to you—"

"No." She shook her head. "This is different." She stared straight into his eyes. "You're different in this one. And the storm, it's different, too." She hesitated again, but in for a penny... "Dr. Masterson postulated the existence of...closed dimensions. Changes that don't spawn new dimensions. One where the changes do ripple forward and change the future. This might be one." She stopped because she was moving into new territory. And he had a lot to take in.

"Masterson?" He swallowed. "He is a dimensional traveler, too?"

"Well, yes and no. He doesn't have the Talent, but he's super smart, so he sometimes travels with one of us." How strange. She'd never wondered when and how he came on board before. He was just there, like her dad.

"How am I different in this dimension?" he asked. He was pale, but that could be the bright light. Maybe.

Oh crap. She closed her eyes for a minute, then opened them. "Ophelia. Her dying is different. In all the others, she doesn't. You're married to her. Or jacketed, I mean, about to get married to her."

His eyes widened. He looked like he might speak, but his lips thinned instead. He walked away from her, his shoulders twitching, then he turned back.

"And the storm? How is it different here?"

"It seems to be less...damaging," she admitted. "But—" she stopped because this wasn't just a wander into the unknown, this was a leap.

"But?" his gaze demanded that she continue.

"Your memory of me, of meeting me. That's really new. I've never, ever...ever had that happen."

"But you have a theory?"

"Well, I'm not sure it's a theory yet. More of a speculation, actually. But the fact that you remember something I don't—I wonder if time is...replaying...or...re-setting."

He started to shake his head but paused. "Re-setting. As if we got it wrong and must make it right."

She nodded, her turn to be bemused. "Is that what you feel when the memories happen?"

His turn to nod. "It is."

"So if we got it wrong...before..."

"How do we get it right?" he finished.

ERNEST PACED BACK and forth for what seemed like a long time. Finally, he turned to her. "If the meteorite in the Blizablighter 3000 is somehow the issue, then I should disassemble the machine."

Prudence started to nod but hesitated. "Does that feel... right?"

It was his turn to pause and consider. His shoulders lifted some. "It feels new."

"Okay, well at least new is different from wrong."

"But it is not known to be right." He rubbed his face impatiently.

He should be used to uncertainty. He was a scientist, a weather scientist. And she really shouldn't have to point out this obvious, but he had taken a few hits in the last minute. "If we've been getting it wrong, then we don't know what's right. All we've got to go on is new."

He stared at her and finally nodded. "You are right." He blinked at that.

"It's ironic, isn't it?" She grinned, and he grinned back, though it was one with lots of strain in it. Wow, three in a row. They were on a roll. They started toward the stairs, but Prudence paused to ask, "Lord Henry? Is he new or wrong?"

Ernest stopped, considering. "I only started to have the... sensations...when we were walking. I do not think—I cannot say."

"Fair enough." She headed down, her booted feet ringing on the metal steps as she circled down to the watch room.

Ernest went to his Blizablighter 3000—where had he come up with that name?—and crouched down. He reached for the wires and stopped.

"I did this before."

"So wrong move," Prudence said, sort of asking, but not really. She rubbed her face. A lot of inter-dimensional study involved feeling one's way in the dark, but this certainly had a new, high level of difficulty. "It's not very big, maybe we can, I don't know, I hide it?" She looked around. There really wasn't a place that a determined search wouldn't reveal. But there were some nooks and crannies around that could buy them some time. Maybe.

Ernest knelt, not talking, his expression intent. Finally, he murmured, "Something about the lamp."

"The Fresnel lamp?" It was designed to amplify light, not electricity, but— "Okay. Where do we put it?"

His head tilted to one side, his glance moving from her to the automatons, then finally back to her. His gaze holding hers, he said, "No." He glanced around. "I'll take Winston and Fred with me."

She nodded, not sure she agreed with him. She angled her Raijin-ator 5 casually, but in a direction where it would be easy to target Niles if he made a move. She flipped it to energy charge. She'd seen how fast he could move. She'd get one shot at him. Problem was, it took multiple bullets to take out an automaton. She had to hope a surge would short out his circuits.

She shifted from one foot to the other, wondering why she felt restless and uneasy. Well, beyond the weird storm and dragons. Metal hit metal above her. She didn't like not seeing what was going on up there. She didn't have Ernest's other sense, so this wrong feeling was probably just about being scared to her toenails. She wanted to look up, even though she couldn't see anything. So she looked down instead. Winston was a dim figure on the last half landing. After that point, the stairs made a complete circuit of the lighthouse before reaching the next landing. He didn't seem to be moving, but she thought she saw the slight glow of his eyes. That was a bit creepy. Her dimension hadn't really gone heavy into automatons when the difference engine was discovered. And it was too easy to mess with their programming, so they weren't used at all inside the Agency. There was no question the ability to travel through dimensions had accelerated the development of other technology available to the Agency, but it was all very carefully controlled. No one wanted to be the CEO who created a catastrophic paradox.

She shifted again, as things grew quieter up top. She wasn't sure she trusted any of the automatons. All four automatons had been out of Ernest's sight at some point, but only the two ship handlers had been outside the umbrella of light. The Storm had appeared during the day, so it wasn't the light that kept it at bay. Was there something about lighthouse lamps? The light they generated that was protecting them? Or was the light an illusion of safety?

She realized even the small sounds had stopped up in the lamp room. What if the automatons had done something to Ernest? There was the clump as first one, then the other descended. But not Ernest.

"Ernest? Are you all right?" She raised her voice to be heard over the footsteps.

"I'll be right down."

His voice sounded fine, a little breathless. The automatons looked the same. Prudence found a niggle of doubt working its way into her certainty that this Ernest was as good and trustworthy as the others. The memory of her missing grandfather surfaced. As far as she knew, they were the two with the strongest Talent ever recorded in Agency history. And he was missing.

What if the trap was not for Ernest, but for her? If someone had been studying her, while she studied the Ernests, how hard would it have been to detect her...lack of distance? The impersonal she'd not been quite able to maintain? All this Ernest had to do was pretend—what had he actually told her? She hadn't told him about Octavius, but if someone had captured her grandfather, they might already know the importance of a familiar. She could transit without him, but without him, she could not transit *home*.

She checked her tracking. So far Octavius seemed okay. As long as her connection was strong enough, they should be able to leave. It was not at the maximum range, but it was close. It was tempting, so tempting the portal started to form in front of her again.

But—what if that was—the desired outcome. What if whatever was out there hoped she'd panic and leave? Or...if this trap was for her, they'd want her to have the illusion she could leave—

The boom of the door far below made her jump. If she hadn't been in a lighthouse, she might have hit the roof. But she couldn't jump that high. Close, but not quite. The adrenalin surge almost sent her into a new dimension, though.

"Dr. Warren?" In cultured accents, Lord Henry's voice echoed up the tower, his voice bouncing eerily off the thick stone walls.

She was not in the mood for him. Was it morning in the time...flux they seemed to be in? It was still dark outside, but that did not mean sunrise wasn't happening. And she had told him he could come back. A soft shuffle came from behind her. She turned. It was Ernest padding carefully and quietly down to her level. His gaze was filled with warning.

She got it, but also felt caught between two unknowns, so instead of meeting his gaze directly, she peeked down. The door was open, letting in some light, though she could not identify its source. It could be the sun or light from the lamp. His dark figure filled the opening, laying a distorted shadow onto the rough floor.

He had to have checked the house first. If he'd parked next to *The Weatherman*, then he knew they weren't there.

"Dr. Warren? Miss Smith?" Some worry filtered into the suave voice now.

Ernest half shrugged and leaned over the railing. "We are up here, sir." He hesitated, then added, "It is a considerable climb."

It was. It gave them a few minutes grace, depending on how fit he was. Prudence hesitated. "I'm going to go look out again." And possibly have a look around the lamp room?

Ernest did not look pleased, but he nodded. "Keep that close." He indicated her Raijin-ator 5.

"Always." Her smile was a little easier, though she could not say exactly why.

He looked at her suddenly. "It was you."

"It was me...what?"

"The name. Blizablighter 3000." He half smiled. "It was you."

Her smile widened. "I hate to admit it, but it does sound like something I'd come up with." Both the Paraton-ator 20 and the Raijin-ator 5 were unofficial names that had somehow managed to filter through the ranks of the Agency. For some reason, knowing this helped. Her breath seemed to come easier, though doubts tried to peck at this sprout of hope. "I won't be long."

She started up. The doubts lingered, but with less force. Maybe she'd rather believe than not. Besides, if this was a trap, well, the fly never escaped by struggling.

UP TOP, Prudence, with her goggles on the active setting, walked all the way around the lamp room. No readings from the lamp, but—her gaze tracked up. There. She blinked and pulled off the goggles. Then lowered them again and turned them to magnify. Slower this time, she pulled them down. Without the goggles, his mineral was almost invisible tucked up there. It looked like the wires from his Blizablighter 3000 now lead up and out the vents to the lightning rod on the top of the lighthouse.

Had he been following his instincts when he put the Blizablighter 3000 up there, or let his scientist out, hoping to test his device while hiding it in sort of plain sight?

She sighed. It might be brilliant. Or not brilliant. Heat and tired sapped at her strength and muddied her thinking. Not that thinking would help that much since she didn't know much anyway. She clambered out the door, hoping she could pretend it was cooler outside. It was a little better. At least the fuel smell was less. She leaned back, remembering her first transit without a trainer. Just she and Octavius making a leap into the void. She'd had a target, and she'd nailed it. But she'd never forget the first time alone in the space between, wondering where and when they'd land. If they did. They had equipment, but the most essential calculation could only happen in her head. In the mind. Before her Talent manifested, she'd asked Lewis how it happened.

He'd shrugged. "You just feel it. There are lines leading all over, and you pick one and follow it."

"But how do you know it's the right line?"

"You just do." He'd ruffled her hair. "They won't let you and your familiar go solo until you know."

It took skill and training to stick the line. There were so many. So easy to run off course. No one knew why a familiar helped center the mind. There were some theories about animal instincts and such. In the end, it didn't matter. It was a two-part process. First came the Talent, then the familiar found the Talent. If the familiar didn't come, well, there were those who tried anyway. Eventually, they were found, sometimes alive, sometimes in a grave. It was a time thing, not just a dimension thing.

If this was what it felt like to lose the line, well, she did not want to go there. And if she was already there? Well, better not to think about it.

She straightened her spine, then her body, positioned her

goggles, and set them to scan again. She made a circuit around the balcony. The light was warm and comforting. Even in her time, lighthouses were special. Was that what had drawn her here? She did one more round, taking it slower and trying to "see" everything. Maybe the readings would be found, even if—

As far as she could tell, the dragons were gone or flying somewhere else. On the pier side, there was the shadow of an airship tied down on the opposite side of the pier from *The Weatherman*. Lord Henry's ride, she presumed. The night outside the lamp umbrella seemed less dark than earlier. She reached the door again, glanced around, suddenly uneasy in a new way. Amazing she had a new way to be uneasy. She thought she'd reached her upper limit.

"My head hurts," she told the night. Her heart felt a bit battered, too. And the sad part? She was the one who'd given it the swift kick. Ernest hadn't done anything. Yet. She sighed, reluctant to go down and face either man. She closed her eyes, rubbed her face. The waft of cool air felt so good, she sighed with relief this time. Cool turned downright cold against her sweaty skin. How could it be cold? Uneasy ramped up to bad feeling. She needed to lift her lids, but man, she didn't want to.

And she wished she'd kept them closed when the prow of an airship emerged from the darkness just ahead of her. It wasn't the fact that it was airborne or even that it was an airship that she found so, well, terrifying. It was the semi-transparent hull, with small pulses of light running along the beams. And the eyes. Thousands of them. Glowing red.

Yeah, those were terrifying, too.

Chapter Twelve

ERNEST WATCHED Lord Henry climb the last flight of stairs. The man showed no sign of distress or breathlessness. In fact, he might have been strolling on a level promenade. His cane swung lightly in one hand. His hat held in the crook of his other arm. There was a regular cadence to the sound of his steps against metal. Like a metronome. Or an automaton.

Too real.

His mind rejected the connection. He looked human. He had skin and normal eyes. He was a man, except, no man could walk with that kind of precision, at least, could soldiers? He did not know. Was that precise timing possible on a rising stairway? The distance between the stairs would have to be close to precise. Even a small difference would throw off the timing.

Lord Henry reached his level and stopped, his chest rising and falling without distress of any kind.

Automatons didn't breathe. Their chests didn't rise or fall. They didn't have lungs. Their clockwork hearts took up most of the chest cavity. There was sweat on his skin. Lines around the eyes. He was a man, therefore, that kind of precision must be possible. At least it was not an impossible possible this time. It only felt like it was.

Lord Henry shifted his hat to a position under the arm holding his cane and bowed very slightly. "Miss...Smith not here?"

Why the pause before her name? "She's servicing the lamps inside the Fresnel."

His gaze flicked up. Automaton eyes couldn't do that. They couldn't show emotion. Not that Ernest could see an abundance of emotion in his eyes, but there was intelligence. Awareness. His eyes were quite compelling, he realized, and found it took effort to look away.

"I'm surprised you made it back, sir," Ernest said. "The storm seems to have reached this location. Or is imminent. We have already experienced some flooding."

The dark brows rose. "But I have only been gone an hour."

An hour? Ernest blinked.

"We started for the mainland, and the wind came up, and we decided to return here. Without the lamp, we would not have found our way back."

We? "Where is your crew, sir. Do they need—"

"My automatons?" Lord Henry looked surprised. "They remain with the ship." His glance flicked toward Ernest's crew, then back to Ernest, a question in them.

His lips tightened, but he did not respond to the unspoken question, nor would he have had the question spoken. He wanted to protest Lord Henry's callousness but was aware of his own hypocrisy. He had, several times, put the automatons between himself and danger, but he would never leave them behind when a choice was offered. Though in this circumstance, it was hard to make the case that they were safer here with him.

"Niles?"

The automaton's eyes flickered to active. "Yes, sir?"

"Go down to Crispin, and both of you keep a watch on the

water levels out there. Oh, and Niles," the automaton looked back, "try to stay dry. I mean, move higher if you need to."

"Yes, sir."

Niles clunked ponderously down the stairs towards his counterpart. As it approached, Winston came back to awareness. For just a moment, it seemed its eyes glowed red, but it must have been a trick of the light. Had to have been because it was impossible. The metal eyes could only be silver. And inside his head, he heard her voice saying, "...six impossible things..."

Had they reached six yet? He was getting rather tired of the impossible.

Now he looked at Lord Henry, his mind returning to his statement that he'd only been gone an hour. Since his watch was not tracking time in a reasonable manner, he had only his senses to track the passage of time. It felt longer, but also shorter. In fact, it felt as if time were speeding up after going very slowly.

"I'm not sure you made the better choice in returning here, sir. If my..." he hesitated, "...instruments are in any way accurate, we are in the direct path of this storm."

Lord Henry looked more pleased than worried. "Do you think it is The Storm?"

"I do not know."

"May I see your instruments?"

"Of course, sir." They were only on the other side of the central pillar, so Ernest led him there. There was a long silence while Lord Henry looked at each one. Finally, he glanced at the crates that the automatons hadn't unpacked. They were mostly papers, but also pieces of instruments Ernest was working on. It almost seemed that Lord Henry looked disappointed.

"Is something wrong, sir?"

Lord Henry's lips twisted. "I had hoped to see something... original from an inventor such as yourself."

He couldn't know about the Blizablighter 3000, now hidden

up in the peak of the lamp room's ceiling. No one had known until Prudence.

"I have some ideas sketched out, but sometimes one can't improve on what has already been done." He half shrugged, hoped he managed to sound rueful. "Indeed, I'm glad to be spared the embarrassment, since none of these are working properly."

Lord Henry looked disconcerted. Had he not known how to read the instruments? He recovered very quickly.

"Perhaps you...picked something up on your travels that is interfering with them?"

Lord Henry knew about the meteorite. Ernest felt it, though he'd have discounted it prior to meeting Prudence. Had he spoken with the driller who gave it to Ernest? That is the only way he could have known. But why would the man have told him? Why would Lord Henry care? What was his interest? Other than the fact that it was not of this world, it seemed a side issue in the face of The Storm. All he could think to do was play along.

Ernest looked thoughtful, nodded slowly. "I suppose that is possible. Winston, when you and Fred packed these crates, did they have any mineral substances in them?" He knew it was a safe question for his very literal automatons.

"The crates contain your research materials, sir," Winston said. "Books and papers."

Fred pointed to one half empty crate. "That one has a few pieces of incomplete instruments that you were working on."

"May I?" Lord Henry asked, already moving toward that crate. He bent over it, grabbing at the pieces of straw in a manner that was borderline frantic. He stopped and looked at Ernest, his smile forced. "These devices look...intriguing, Doctor."

"Well, so far they have been frustrating." Ernest picked up a piece of one, turned it in his hand, then dropped it back in the crate.

Lord Henry glanced at the automatons, then the stairs. With

lowered voice, he said, "I've come with a warning for you, Dr. Warren. From Dr. Masterson."

Prudence had come from Masterson. Or said she had. But he'd felt that the knowing he'd felt about her was real, that this sense of doing this before—a knowing that was *not* knowledge, he reminded himself. A trap? With her up there and Lord Henry here? He had only their word they didn't know each other. And her wild story. And honest eyes. Only that and his instincts to put against wealth and eyes that were...eerily compelling. And as calm as his breathing. *Too real?*

"You are in great danger." His gaze flicked toward the stairs again, as if to indicate Prudence was that danger.

There was a clatter of some sort overhead.

"Ernest?" she called out, "Could you step up here for a minute?"

Her tone was in opposition to her words, he noted, wondering if the trap were closing. And who was supposed to be in it?

"Take care," Lord Henry cautioned, grasping his arm with a cold hand.

He should have had that advice four years ago, Ernest thought somewhat ruefully. Despite his concerns, Ernest shook off Lord Henry's hand, and went for the stairs, felt cold air spilling down from the lamp room, as he leapt up two at a time. How could a room that had been stifling and hot now be cold as the arctic?

Prudence stood at the top, gripping the rail, her face so white her eyes looked like black stones.

"What's happened?" The chill was so intense, he could see his breath and hers, like fog filtered with the lamp light. He heard Lord Henry climbing up after him and grabbed her shoulders, had to tug to get her to let loose the rail. She was cold, even through her leathers. His hands slid down her arms and gripped the chunks of ice that were her hands.

"What's wrong?" he asked again. Instead of answering her gaze tracked past his shoulder. He followed her gaze, half turning them both in the process—

This was one impossible too far. And yet—hadn't he already seen this? Just before he blacked out on *The Weatherman.* "Wait here," he said, his grip loosening, but her hands tightened.

"No." Pale tongue licked dry, barely visible lips. "I...I'll go with you."

The instinct to protect was strong, but she unslung her weapon, holding it with the muzzle pointed down. He opened the door. It wasn't on the latch, so he only had to pull it back. He bent and stepped out on the ledge. It had seemed generous earlier. Now, it did not.

A raucous cry, quickly followed by that of its mate, sounded overhead. He was somewhat aware that the dragons were back, that they circled above the light, shadows weaving past each other in a regular pattern. The ship was a disturbing sight. Transparent, with pulses of light tracing along its structure. There was fog, but mostly in wisps that clung to the sides of the airship. A bit of fog would push out and then jerk back, as if it hit a barrier or worse.

And there were the eyes. He remembered the eyes. And this cold, he realized. He'd felt this cold during his last encounter with this ship. He wanted to consider the temperature because it was easier to contemplate than the eyes. So many of them, staring at them. Eyes that glittered. Eyes that pleaded. Eyes filled with fear. Eyes filled hatred. He searched them, afraid to see, and afraid to not see, her eyes.

He didn't realize he'd relaxed some, until Prudence spoke, her voice hoarse. "She's not there."

He wrenched his gaze away, looked down at her. If anything, her skin was whiter than it had been, and tightly stretched across her cheek bones. She shrank back as Lord Henry climbed out next to her. He looked...curious.

"So these must be the specters in the storm," he said. He

looked at Prudence, his mouth acquiring a cruel edge. "Do you recognize anyone?"

There was a long silence as she stared at him, then imperceptibly at first, her back straightened, then her whole body. She pushed away from the lighthouse until she stood on both feet. She opened her mouth, but before she could speak, there was a rattle and groan, though not exactly a groan either. The sound was unearthly.

Ernest felt such dread, it was all he could do to maintain his stance. Prudence's grip on his hand tightened. The visible specters appeared to be falling back, or to either side, clearing a path if it was possible to have a path over nothing. Some of the specters looked as if they cowered back, while others had gloating looks in their red eyes.

"What is it?" she asked, looking to Ernest.

But it was Lord Henry who spoke. "Don't you recognize him? Granted he has changed..."

WHAT WAS HE TALKING ABOUT? Recognize who? Prudence felt both slow and stupid and scared silly. But the warmth of Ernest's hand kept her grounded, well, mostly. If this were the trap, she'd have preferred a spider web. Probably. She didn't like spiders either.

A figure began to emerge from the inky blackness at the back of the ship, a gray human-like shadow that moved slowly, each step sending a ghostly rattle that made some of the specters crouch in fear. Others seemed to laugh, black hole mouths opening beneath those red eyes.

In the air in front of her, a portal tried to form. Just a shimmer because she fought it back. She didn't look at Lord Henry, but she felt his pleasure at this small betrayal. The figure reached the prow of the ship. His essence seemed to quiver, then sharpen until she could see it was a man. A man in chains, chains that glowed briefly blue with each step, chains wrapped around him from the top of his head to the feet visible through the transparent side of the ship. Only his face was free enough to see. His eyes were wild and tormented. So much pain in them...the weight of his pain dragged her down. Her knees started to buckle, but Ernest grabbed her, his arm sliding around her waist to keep her on her feet.

Those eyes turned slowly toward her. She wanted to cower from them, from discovery, but she couldn't move. The eyes saw her. They showed recognition. How was that possible? The mouth opened, releasing a scream that reverberated through her skin and bone. And through her mind. With rising terror, she saw fine cracks in the dimensional lines inside her head, saw some of those lines get damaged, some disconnect. If one of those was the one home...

She felt herself being dragged back, away from the chained man, away from the ship, and then she couldn't see him with her eyes, but it didn't matter. The eyes were in her head, looking at her through the lines. The broken lines. She couldn't—and then it hit her, and she knew. Her head fell back. Against a shoulder, that felt so safe and warm, but safe was an illusion. A ruse.

"What's wrong? What has happened?"

Ernest sounded frantic. Not like himself at all. And so very far away. With some effort, she turned her head toward him. Saw panic in his eyes.

"What was it?" He gave her a small shake.

"Not what. Who. My grandfather."

"Your grandfather? What—" The color drained from his face. "That's him...but...why?"

"I'd rather die..." she said, surprised at how loud the words sounded in the deep silence.

He pulled her to his chest, holding her so tight she almost couldn't breathe. She didn't care. His heart beat against her cheek. He was alive. She was still alive. She wasn't sure that was good.

"You're not going to die," he said. "I—"

She grabbed his jacket lapel. "Promise me you won't let it have me. That you'll kill me rather than let that happen."

"Pru—"

"Promise me you'll use the gun."

His lips thinned, then he set his mouth against her ear. "What if that is what we did before?"

This question cleared some of the panic. She took several breaths. "Does it feel wrong?"

He hesitated. "Of course it does. But I can't tell if—"

"Dr. Warren?"

She tightened her grip. "I don't like him."

His lips widened into the strained version of a smile. "Neither do I."

"Can you help me up?" She honestly wasn't sure she could stand, or stay there if he did help her up, but she wasn't going to cower at the feet of Lord Henry or—whatever it was holding her grandfather prisoner out there. He was a prisoner. The chains confirmed that truth. She managed to get up...just. She gripped his shoulders. "I just want you to know, no matter what—"

She realized he wasn't looking at her, but past her. *I don't want to know.* Despite this, her body somehow turned to look. Her grandfather was still there, but he'd been joined by someone. A woman.

Prudence felt Ernest stiffen as he recognized her, too.

Ophelia.

"AND NOW YOU ARE FINALLY...FINALLY beginning to understand," Lord Henry said, his tone one who had waited far too long for this moment and was determined to enjoy it to the maximum.

"And yet somehow I don't," Prudence said, her voice thin and strained. "What's wrong with him, with...them?"

"There is nothing wrong with them. They rejoice because their life force serves the Master."

"Yeah, they looked like its the jets. Thrilled to their toes—oh wait, they don't seem to have toes."

Ernest found he could almost smile, even though some of her words puzzled him.

There was a flicker, then the lighthouse lamp lost some of its brightness.

"One of the lamps inside the Fresnel has gone out," Prudence said, almost absently. "I should refill it."

If she felt like Ernest did right now, she could not move. The ice in the air had moved into his bones, into his sinews. As if the lessening of the light allowed it, more of the airship moved closer to the lighthouse tower. One of the dragons dropped through the light canopy, landing on the railing of the balcony. Its red eyes regarded them through the glass, then blew a stream of fire against it. The room grew colder still.

"It's an automaton," he murmured, not sure he'd said the words out loud until Prudence stirred in his hold. "But also more..."

He stood between her and Lord Henry, who held the door open with his body. The dragon dropped down on the ledge and started for the door, its pulsing red eyes seemingly fixed on Ernest. Each step shook the lighthouse tower. The steps seemed to reverberate inside the storm, too, like echoes of distant thunder.

"If you have it, Dr. Warren, now would be a good time to produce it," Lord Henry said.

The dragon was more clumsy on the ledge, but it hadn't far to go to reach the door. It filled the doorway, flexing its wings once and crying out. Its head reared back, the neck elongating as if it was preparing to spew fire, but a light shot from beside Ernest before the dragon could spew. It tracked toward the dragon, straight and true and cleanly white, striking the dragon in the chest.

Lightning danced across its metal surface, giving flashed views of its mechanical insides and then it tumbled back, crashing through the railing and falling out of sight without a sound.

What the—he looked around and realized Prudence had managed to raise her long weapon and fire.

Lord Henry looked at Ernest, his eyes suddenly turning the same red as the dragon.

"That was unfortunate."

"Maybe," Prudence said, "but it felt good."

From deep within the storm, thunder sounded louder and closer, like a death drum, rising and falling.

"You've made him angry." Lord Henry smiled, his eyes gleaming as if anger pleased him.

Ernest stepped more fully between Prudence and Lord Henry. He laughed. The sound of it echoed again and again, building instead of diminishing.

"You cannot help her," he told Ernest. He lifted his cane, exposing the head. It was a smaller version of the dragon Prudence had shot. Its eyes were red, too, and its mouth opened to expel a smaller spout of flame.

Then the screaming started. Thousands of voices, as if the shot had hurt them, too.

Chapter Thirteen

IT WAS hard to imagine what could be worse, but Prudence had a feeling that worse was still incoming. And then, as if the worst weren't bad enough, another oil lamp went out inside the Fresnel. The lamp's light dimmed some more, and the airship nudged closer, angling so that a ramp—one-half wood and half fog—could be lowered to rest in the place where the railing had broken. The brimstone smell intensified, but so did the cold.

And then the screaming cut off as if someone had flipped a switch.

She felt pinned in place by Lord Henry's creepy human eyes and his dragon head cane. Off to the left, her grandfather's specter turned to face them, his burning gaze both entreating and filled with warning. Not that she needed the warning. The urge to flee was so strong, it took all her focus to keep the portal at bay. Her gear showed a strong connection to Octavius. Ernest gripped her hand. Emergency transit might save them, if her gear was accurate. Had her grandfather faced this horror? He would have followed protocol and tried to transit. Only here he was. Caught like the fly in the creepy web. So that seemed to rule out an emergency transit. Which meant she probably couldn't trust the readings. They might be as messed up as Ernest's instruments.

Was it possible to craft a dimensional transit trap? She glanced at Lord Henry and caught him licking his creepy thin lips in anticipation. He loved what was happening. Couldn't wait for what he thought she'd do.

The analogy of the fly and the spider kept rising in a brain almost frozen with panic. But not quite.

Flight or fight.

If flight was out and fighting was indicated, how did one fight specters? Ghosts? A life force-collecting Master—

The boom of footsteps against ghostly wood yanked her attention back to the ghost ship. Amidst the drifting Fog or mist, she thought she saw a figure. Slowly, it emerged from the mist, stopping at the top of the gangplank. Ernest inhaled sharply. Prudence more slowly because her chest was so tight breathing was only barely possible.

It was a specter-distorted-mirror of Lord Henry.

All she could say about him was that he made the human Lord Henry look good. All that was hideous about the man was worse in the specter. And his skeleton was visible through a surface as transparent as his ship. The clothes were tattered and floated around him like the fog clinging to the ship. He didn't doff his ragged top hat. Apparently, manners weren't as important after... specter-fying. He leaned on a cane that was a complete twin of the one held by the human Lord Henry. It sparked when the tip touched the gang plank. Through his fingers, red eyes glowed, from what she presumed was another dragon head.

She'd never liked horror movies. If she managed to survive this, she'd be a lot less judgmental about what characters did while in them. She knew her thoughts were silly, random even. If she focused on stupid stuff, maybe she could keep from panicking. Because if she panicked, well, she had a feeling that is exactly what they wanted. A fly struggling in this terrifying web. She didn't know how she knew this, and technically she didn't *know* it, but it felt true. If the human-ish Lord Henry liked panic and pain, then how much more would his creepy counterpart be panting for it? Her gaze flicked back to her grandfather. To the chains that contained him. Why was he the only one in chains? Why...no, the question wasn't why. It was obvious why he'd been caught and chained.

He was an inter-dimensional Agent. They were using him—the man with the most Talent ever—to travel between dimensions. To move more than a few people, but a ship. Okay, it was a specter ship, but there was the storm. That's how it could move between dimensions, how it could appear and disappear. That's how he'd been able to play the Dr. Masterson card. Though it didn't explain why it was eating other dimensions—something to do with the life force it required, maybe? Her grandfather's specter looked less than the others. Had they sucked the life out of him and were hoping she could take up the job? Did she have enough Talent to move something that large between dimensions? She wasn't sure she knew how—

The whys and hows didn't matter. They now had the only other Agent as Talented as her grandfather trapped—or almost trapped—in their web. The distinction was small but necessary to keep her from having a hysterical fit.

She glanced at the human-ish Lord Henry. Why two of them? Why—Ernest? And Ophelia? What did he have—

Lord Henry had funded the geologic expedition. While she'd been up in the lamp room trying to get her voice back, trying not to scream like a little girl, she'd heard him ask Ernest if "something he'd picked up in his travels" was causing the problems with his instruments. The only item that fit, and was related to that contact, was the meteorite. Did they want it? Or did they fear it? Or was it both? It collected and stored energy. Might a life force be an energy source, too? Could they collect peoples' souls, their life force, with the meteorite? Or did they hope Ernest could do it for them? Was this trap for both of them?

Lightning began to build around the ghost airship, emanating from the hull, or so it seemed, dancing across the sky and hitting the lightning rod at the top of the lighthouse. She studied the flashes and realized that just as she could see inter-dimensional lines in her head, there were connected lines in the seemingly random flashes. All of them leading from the lighthouse to the ship. Or from the ship to the lighthouse. It was hard to tell. But they were connected flashes, not disconnected flashes.

It was doing something. She wished she knew what, because she had a feeling that it needed to be stopped. She had an idea. Or the ghost of one.

"Doctor Ernest Warren." The specter's mouth opened, but the words seemed to come from the human Lord Henry.

Ernest's grip on her hand tightened. Might have been painful if she'd had feeling left in it. But it was so cold, her breath was a blue mist hanging in the air in front of her.

"I am here," he finally said, his tone so calm, she'd have been impressed if he wasn't smashing her fingers to pulp. "What do you want?"

The red gaze fixed on Ernest, then traveled to Prudence. She lifted her chin, let defiance show in her eyes. She hoped.

"I want the witch."

Witch? Prudence's brows rose. She'd have been offended if she wasn't so busy being terrified.

Ernest's mouth opened, but the Lord Henry specter held up a claw-like hand, the one holding the cane, and pointed the dragon head at Ophelia. Red light shot from it, circling her like rope and dragged her to the top of the ramp.

"Ernest," she cried. "Help me!" She writhed in its hold as if the rope hurt her.

Hard to believe he could, but Ernest managed to stiffen some more next to her.

Prudence managed to stiffen some, too, as she sensed what was coming.

"Your true love," his voice was thick with derision, "for the witch."

Yeah, not a surprise.

Your true love for the witch. The words echoed inside Ernest's head. For a moment, his heart rebounded with joy at the thought —but reason quickly killed the hope. That wasn't, that could not be his Ophelia. And how could he hand Prudence over to that monstrous thing? How could he trade a lost life for a true life?

"It's a lie," he heard himself say. "You lie. You can't resurrect the dead. It's not possible." Even as he said the words, he heard Prudence's voice in his head. *Six impossible things.*

"I am proof it is possible," Lord Henry spoke next to him, using his voice, not that of the specter. "I am real. Living and breathing. I died," his red gaze flicked balefully toward Prudence, as if he somehow held her responsible, "and came back better than before. She will return, even better than your memory of her. You can have your love, return to your real life, and forget this ever happened. If you give us the witch."

He probed his senses, hoping for something, some memory of what they'd done before to guide him now. But all he felt was horror. Horror at the thought of Prudence in the clutches of that thing. Horror for Ophelia, trapped there. How could he choose? But how to fight ghosts? Specters? Her weapon had destroyed the automaton dragon, and perhaps could kill Lord Henry, but if that thing could bring him back—

"Do it," Prudence said.

He jerked, turned to stare at her. She looked at the specter, not at him.

"Take the deal."

"No." The words were forcefully uttered, so forcefully it shocked him. The specter next to Lord Henry was not his Ophelia and she wasn't his true love. It was a bad time to realize she probably never had been. Oh, had she lived, he would have been faithful, but what he felt for her—he stopped the thought. He had to stop it. Because if he admitted it, if only to himself, then he'd have to take the deal. From guilt. Not because he loved Ophelia so much, but because he did not love her enough.

"Yes." She angled toward him, her smile beyond strained. Her eyes were bruised circles in her white face. "There are so many reasons not to," she managed a sad chuckle, "*three thousand* of them if we had time to count," her gaze flicked down to his waistband, then back to his face, "and not one good reason to do it. But you need to make the trade. You should save Ophelia." And now her expression was wry, but there was warning and something more in her gaze. "The fly can't escape the trap. You might as well save one."

"She speaks truth," Lord Henry said, amused. "Unusual for a witch."

Her hand seemed to touch him, but instead, she did something to the weapon she'd given him, the small click lost in the booming sound of the specter's gloating laugh.

She squeezed his hand one last time, then dragged her hand free from his clasp, from his hand that wanted to cling, to hold on to her and never let her go.

"I—" he stopped. What could he say? He nodded, to let her know he'd received her message.

"It's all right." She leaned in to press her lips to his cheek, whispered in his ear, "When I reach the top of the ramp." She turned from him to face the specter, ignoring the human Lord Henry. She leaned her long gun against the railing and straightened her shoulders. "Send her down."

The cane head pointed at Ophelia again, the red rope falling away.

"Ernest," she said, starting toward him with her arms out, remnants of flesh hanging from the bones.

All he felt was horror and dread. The dark maw of her mouth, the merciless gaze. The red eyes. He fingered the weapon. He had good aim, even under duress, but never had he needed accuracy so desperately. He did not expect the apparently human Lord Henry to give him more than a shot or two at the Blizablighter 3000. Was this why it had felt right to place it up there? Or was he about to get it wrong again? A shadow moved across the base of the airship. The last dragon, he realized. It would attack, too, most likely. In any case, he'd rather reset time and do this again than live his life with that travesty of Ophelia.

PRUDENCE DIDN'T RUSH toward her fate. One didn't. Not if one had even a smidgeon of survival instinct. In fact, it took all her resolution to step onto that semi-transparent gang plank. She lifted a booted foot and tested it. It held, without feeling solid. She glanced across, met her grandfather's gaze, saw his slight nod. If his free will hadn't been completely subdued, it was a good sign she'd made the right call. Funny how that didn't help all that much. And she didn't know if his free will had survived. He might be part of the trap. Maybe they didn't care if the fly came willingly.

Prudence brought her other foot onto the gang plank. She seemed to sink in an inch or so. She tried to lift her foot, which shifted her weight back and the misty goop tightened around her feet. She couldn't move until she leaned forward, could only go forward, she realized. She leaned forward and was able to take a couple of steps, each time sinking deeper into the goo. Hopefully, she wouldn't sink so deep she'd fall out the other side. It was a long drop.

Ophelia didn't rush either. She sashayed, which was not pretty since she was mostly skeletal. The remnant of her lips curved in triumph, her gaze dismissed Prudence, though Prudence wasn't sure if the dismissal was as a woman or as a threat. If the gal had had access to a mirror, she might not be quite so confident. Because Prudence was a woman and Ophelia was only sort of a woman, Prudence arched her brows returned the dismissive look, then rolled her eyes.

Ophelia's hands curled into claws and she made a move toward Prudence, but a blast of cold air seemed to push her down the ramp. They passed each other—holy cow the smell—without contact. And, weirdly enough, Prudence began to slide up the gang plank without any effort on her part. Like she was on a conveyor belt now, moving toward her nasty fate.

She looked down, trying to see if there was a mechanism of some sort. That's when she saw the fog creeping up the sides of the airship, and then across the gang plank toward her. The flight instinct surged as the white mist crept closer. It took all her willpower to hold the portal at bay. Like icy fingers, the fog wrapped first one ankle, then the other, then it began to wind up both legs. At first, it was so cold it ached, a cold burn, even through the boots. She thought she heard a soft clink and felt the additional chill of metal against bare skin, sinking into her bones...

Now panic clawed at her throat. A scream tried to push out. She wouldn't let it, though she choked on it. She felt Octavius's panic. Below her, she saw the dragon circling the water, shooting flames at the surface as if daring him to come out. She couldn't help him. Couldn't help herself. If Ernest—even if he hit the mark, would it work?

Just because the fly didn't struggle, didn't mean it wasn't trapped. That it could escape.

Her hands fisted at her sides as the fog tightened its grip. The burn was just short of intolerable. She shuddered to hold in the moan. Around her the specters moaned for her, some with pleasure, others in what appeared to be shared pain.

And she drew inexorably closer and closer to the Lord Henry specter.

Like his human counterpart, a tongue snaked out, licking the remnants of lips.

The closer she got, the worse the stench. That's what she smelled, to a lesser degree when close to the human version, she realized. What did that mean? Was a part of him still dead?

The eyes, the other specters behind those eyes, were clustered on the deck, a well-defined gap between them and specter Lord Henry.

The Lord Henry specter smiled, the teeth in his visible skull parting in a rictus grin. He lifted the cane head, moving it in a circle that wrapped her in the red ropes that burned a new way, and turned her like a puppet. At his side, but facing the lighthouse. Facing Ernest and the avid human Lord Henry. A hand, more bone than flesh stroked her hair and cheek, and this time a moan escaped her lips. Despair crept toward her heart, its fingers as icy cold as the specter's touch.

"So warm," the specter purred, leaning close and inhaling—the action and words echoed by the human Lord Henry. His tongue snaked around his mouth again, and the remnants of lips curved up. "So sweet. And all mine. I look forward to introducing you to my ship."

Her stomach roiled, but oddly enough, that wasn't the worst pain she felt. They wanted her to see Ernest and Ophelia. They wanted Ernest to see her. They wanted both of them to lose all hope. That was the power of the lighthouse. Hope. A faint flicker of it pulsed in her heart.

Specter Lord Henry and human Lord Henry lifted their cane heads, directing red light at Ophelia. It struck her like a blow that made her scream, but then flesh and skin flowed in a slow wave up from her feet and across her body. She gasped as her lungs breathed again, her shoulders rising and falling. Her hair, the tatters of blonde turned into a liquid gold flow.

Prudence couldn't see her face.

She could see Ernest's.

His eyes widened with shock, and it seemed his hand slackened on the trigger of the weapon that his hand hid from them.

"Ernest." The voice was Ophelia's, richly caressing, filled with charm. "My dearest love..."

Prudence felt that hope flicker waver—and the chains tightened so that she couldn't speak, couldn't breathe.

Darkness began to close in. The ramp turned into a long tunnel with Ernest at the end and the golden Ophelia moving to block her view of him. And to block his view of Prudence.

"I've missed you so much..."

ERNEST COULD NOT BELIEVE what he saw. It was Ophelia, one better than his poor memory. How could he have forgotten her? Had he really thought he was over her? She was perfect. She was real, so real, he only had to reach out and draw her into his empty arms...

So real.

He drew in an unsteady breath. Would he have forgotten this perfection? It was not easy when his thoughts and feelings were clouded with hope and fear, but he tried to match this reality to memory. And as he did, Lord Henry's words came back to mock him. *I am better than before.*

Too real. Too good to be true.

Relief swept through him as his mind cleared. His gaze shifted to Prudence, slowly disappearing into the circling fog. A fog wrapping her in chains, he realized with horror as the fog shifted, exposing legs beginning to fade like the other specters...

The specter Lord Henry stroked her cheek, his eyes gleaming with perverse possessiveness.

"She's mine now," he said, his mouth forming words that came out of the human Lord Henry's mouth.

In one frantic motion, he pulled the weapon she'd given him and pointed it at the shadow that was his Blizablighter 3000 up in the ceiling's peak, pointed it at the meteorite hoping against the overwhelming despair that he wasn't too late. The pure white light struck the meteorite, or seemed to. He fired again. And again.

He heard the sounds of a struggle below and looked down. Saw Niles and Crispin fighting Winston and Fred in the stairwell. Niles ripped Winston's head off and tossed it aside, then tossed the headless body down the stairwell, too. He looked up and started up the stairs, his arms out. Crispin and Fred appeared to be more evenly matched. He had no time to watch or help him as Lord Henry closed on his other side.

Ernest fired at Lord Henry, then fired again. Light danced across the surface of his body, giving glimpses of bone and a difference engine brain inside his skull.

Overhead, the meteorite flared with gold and white light, veins in the stone picking up that light and carrying it up the wires he'd attached to the lightning rod. The light flashed around the lamp room as well. But most of the energy appeared to surge in a series of masses back along the lightning toward the ghost ship.

One of the flashes inside the room struck Lord Henry lighting him up again. His body shuddered, and he shrieked with rage as his human shell began to disintegrate, the pieces rising and spinning as if they'd been picked up by a small cyclone. With a final surge of energy, fire and smoke consumed him in a bright, red flash.

The specter Lord Henry shrieked in rage as the bright white light continued to build out from the lighthouse, sending more and more masses of light along the connected lightning flashes and back to the airship, hitting it like multiple canon shots.

Ophelia screamed as one of the masses of light hit her, and she vanished in a ball of white and red.

More and more of the light balls slammed into the sides of the airship, punching holes that quickly expanded as if acid ate at it. Blue sky showed through the holes as the storm began to dissipate, too. He caught a glimpse of the morning sun rising in the east.

A massive ball of light, bigger than the rest, surged up the gang plank toward Prudence and the specter Lord Henry, whose mouth gaped in an unearthly howl of fury. He tried to hold onto Prudence, but his arms turned to dust and got caught in a small whirlwind that spun the dust away from her.

As the light ate away the ship, the eyes winked out, like stars vanishing with the rising sun.

More light pierced the darkness, the rays of the sun shooting across the water, joining with the last light from the Fresnel and the "storm."

Ernest saw the gangplank beginning to erode away from the lighthouse. In mere seconds Prudence would be stranded over nothing.

"No!" He started toward her, but there was no way to reach her.

The last bit of the specter Lord Henry laughed, the sound choking off as his head turned to dust and then the airship, all of it, including the gangplank, vanished. Prudence seemed to hang in the air for a second. He gripped the railing, his hand out as she started to fall. Out of a low cloud, the dragon screamed towards her, his wings back, his mouth open wide...

The sea boiled to life, multiple arms reaching out and grabbing it in mid-flight, pulling it down into the water. Steam rose briefly from the spot where, a few seconds later, Prudence hit the surface and sank out of sight.

As if releasing one, last act of malice, a shockwave hit Ernest, throwing him backwards into the lighthouse.

Chapter Fourteen

PRUDENCE TUMBLED HELPLESSLY AT FIRST, head over heels, the water rushing toward her. From that height, if she belly flopped—with some effort, she got her feet pointed down. Saw the dragon heading toward her—and Octavius grab it. Had time for a smile as it got yanked into the water, and then she plunged in after it.

It knocked the breath out of her. Felt like she went down and down forever, so far she'd never be able to get back to the surface in time. Did she see Octavius struggle with the dragon? She wasn't sure. It was so strange under the water, like a dream and yet—her arms felt heavy and her clothes added to the drag. She managed to raise her arms, though it felt like slow motion. She cut through the water with a weak breast stroke, hoping to stop the descent. Gradually, but far too slowly, she began to rise. Her lungs ached for air, felt as if they'd explode if she didn't inhale.

Strong arms wrapped around her, no, not arms, tentacles.

Octavius. In his hold, she surged up, barely conscious. They burst clear of the water just as Prudence had to inhale or explode. She sucked in the warm, salt-tanged air with gratitude and more than a little surprise. She leaned her head against his dome. "Thank you, sweet baby. I'm so sorry I got you into this."

A tentacle patted her head, while another held her securely. He started them toward the shore. A figure stood there. Not the one she'd expected.

Octavius lifted her onto a block of granite and drifted back. Her chest still heaved. Water ran down her face, obscuring her vision. Water ran out of clothes that felt like weights wrapping her arms and legs. She peered through the slowly thinning stream of water and managed to scramble into a sitting position.

He was still there.

She honestly didn't trust her eyes, her senses, anything after what had happened.

He worked his way across the blocks with surprising nimbleness, until he reached her, helped her to a more secure spot at the base of the lighthouse. The morning sky was deeply blue but streaked with glorious streaks of purple and blue. As near as she could tell from below, the lighthouse was undamaged. She leaned way back. Even the railing was back to normal, it seemed.

"Why did you jump?" he asked.

She looked at him, this man who looked like her grandfather—a much better version than the chained specter, by the way. She opened her mouth. Closed it. Finally managed, "I didn't jump. I fell."

He looked up, his expression dubious. "Did you?" He looked around, like an absent-minded professor, looking for his glasses. "Well, I'm not quite sure...I daresay we should...this appears to be an island, but I'm sure there's someone..." He stopped, frowning down at her. "Do I know you? Have we met?"

She looked like his wife, or so she'd been told. What she didn't know was how her grandfather could have survived that...whatever it was. If time, or something else, had done a reset, why did she remember—she looked up, where the ghost ship had been and held back a shudder of remembered horror. Would have been very happy to forget that, but not if it meant forgetting Ernest. Before she could speak, he rubbed his face.

"I apologize. That was quite rude. I'm Franklin Pinkerton the Fourth."

Her dad was the fifth. Did he remember her dad? "I'm pleased to meet you, sir. I'm...Prudence Pinkerton."

A frown formed slowly between his brows. "Prudence...I don't—Prudence?"

"Yes, sir."

He studied her curiously. "You have the look of my wife. I thought you might be—but you are too young."

"My father is Franklin the Fifth."

He frowned. "Why don't I remember you?"

"We've never actually met," she said because that was easier than the long answer that he might not remember. If he was back properly, did that mean they'd done it? She looked around, then up. Where was Ernest? "I'm sorry, sir, but there was someone in the lighthouse, I need to check on him, them. I'll be back, just sit here, okay?"

He nodded, seemed almost happy for the time to reflect.

She ran up the steps and inside, but when she looked up the winding stairs, she found she couldn't rush. Dread added weight to her climb. Or maybe gravity was in a bad mood. Around and around and around some more. What would she find up there? If proper time had been restored, maybe Ernest wasn't here. Maybe he was back in Galveston with Ophelia. She saw a window and looked out. Her knees sagged a bit.

The Weatherman. It was still there at the end of the rustic pier. She could see some automations moving around the deck. She reached the watch room. No sign of Ernest's instruments, or the crates of his notes. Were they back on the ship or lost? She heard movement up above her. Was it Ernest up there? And if it was, what would she find in his eyes this time?

ERNEST RUBBED HIS HEAD. It ached so strangely. He explored the source where most of the pain radiated from and found a lump. What on earth? He felt as if he'd wakened too soon from a dream —or a nightmare—but one he hadn't wanted to leave. One he needed to recall.

He touched the back of his head now, though carefully. He remembered arriving here, though that seemed strange and murky in some half-remembered way. The light keeper had let him stay, but why had he come up here? The morning was clear and warm. He inhaled the humid, salty air. He felt...lighter. As if he'd left a burden behind. Was this what they'd meant when they had said he would get over his loss?

He could think of Ophelia without pain. And yet, there was still pain and a feeling of loss. But what else could he have lost?

He heard footsteps on the stairs. Not an automaton, so it must be the light keeper. A woman's footsteps. But how did he know that? He turned from the railing and climbed back inside the lamp room, a strange sensation of hope stirring inside him.

First, the top of a head came into view. Bedraggled, wet hair, then a pale, worried face. Brown eyes, anxiety in them. An uncertain smile flickering on lips...

Those lips. Surely...

I know her, he thought, but how could he? And yet he did. His heart leaped with joy even as his brain struggled to understand.

"Pru—" For a moment he stalled, then, "Prudence?"

A real smile flowed across her face, though some uncertainty remained in her eyes. Worry as well.

She was drenched, trailing water still. Some trickled down the side of her face, and she dashed at it impatiently.

"What do you remember?"

A frown creased his forehead. "You...fell." He jerked around, pieces floating inside his head. Disconnected. Strange. Impossible. *Six impossible things.* "Impossible things?"

Her hands closed around his. That felt right, so right. He

clung because it helped the floating pieces spin slower. Some of them drifted close as if they wanted to fit together.

"I'm afraid there were a few more than six of them," she said, apologetically. "Sorry about that."

He freed a hand to push back a damp strand of her hair. "I don't remember you, but I do. I..." he hesitated, but felt it had to be said before it was too late, "how can I love you and not remember who you are?" His grip tightened on hers. She could vanish. He must keep her here. He frowned. "Do you have a giant octopus?"

She laughed then, the sound as bright and hopeful as the morning. "I do." She sobered. "And I love you, too."

He stared at her, absurdly happy, though he should not be. "Why—"

"I think, I hope," she said, moving closer so that he had to put his arms around her, "that this will help. It did before."

Her chin lifted as she presented her mouth. Before...

He might not remember everything, but he did remember how to kiss a woman...no, how to kiss *this* woman. And he was very eager to do it again...

Chapter Fifteen

"WHAT HAPPENS NOW?" Ernest asked, his fingers entwined with hers and sending tingles up her arm in a way that seriously interfered with her cognitive abilities.

It wasn't easy strolling on Sand Island. So much of it had eroded that they were limited to the boardwalks, which offered no privacy at all. From the *Weatherman* his automatons watched them, and the light keeper and his wife had returned, so she figured they had eyes on them. Octavius and Darby, her grandfather's dolphin familiar, seemed to be watching them. And then there was her grandfather. He sat rocking on the porch of the cottage. The rhythmic back and forth almost hypnotic. Maybe it helped him recover his memory.

It wasn't just memories in pieces. Her drone was gone. Or her equipment was so damaged it couldn't connect with it. That was possible. Her data storage showed empty. And her goggles were just goggles, no sensors or anything, without a drone to connect to. In some ways, it was as if none of it had happened.

And yet here they all were.

And there were signs it had happened, not just her memory of the events. Ernest had a couple of bumps on the back of his head, and she had, well, she wasn't sure what to call the marks on her legs from the ghost chains. They were fading fast and looked like burns, but thankfully they didn't hurt anymore.

"I don't know," she admitted. She didn't. None of this was supposed to happen, particularly the part where Ernest knew who and what she was. For good or ill, she'd changed his life, changed his future in this dimension. It was possible that this had already spawned an alternate universe. She wouldn't know until she returned to the Base. But if she returned with a guy, she was pretty sure they'd notice, even if she also brought back Franklin Pinkerton the fourth, the lost Talent, slash, grandfather.

She'd be out of the Agency. No appeals. What she didn't know is what they'd do about Ernest. If there was a procedure, no one had shared it with her, probably because they knew that knowledge could lead to, um, worse rule-breaking to avoid those consequences. But considering that they'd proposed erasing Ernest as a way to end The Storm, well, going home wasn't at the top of her list. At least, not without a plan.

She touched her temple. And then there was the other question. Could she go home? It had felt like her Talent lines had been damaged when her grandfather screamed, but she wouldn't know until she tried to transit. The reset, or whatever had happened, may have undone the damage, but the marks on her legs said that it hadn't undone everything. The truth was, she was afraid to look.

"What do you fear?" Ernest asked, stopping so that she had to stop and look at him.

What could be worse than what happened, was probably what he meant.

"Losing you," she admitted. "We're from different times, Ernest, different dimensions."

He studied her for several seconds. "You broke the rules that govern your agency?"

She nodded. Though "breaking" was something of an understatement. More like broke them, jumped up and down on them, and set them on fire. She looked past him. The sun was higher now in the eastern sky—she frowned and tipped her head to the side. Ernest's brows arched, and he turned to look, too.

"What on earth—it looks like it is bending?"

"It does," she admitted. She turned in a slow circle. The bend extended all around. In fact, it kind of looked like they were in a... bubble? Was that even possible? And why was she wondering if it was possible when she'd just taken a trip through the crazy impossible?

She realized that Ernest was studying her grandfather, who had risen from the rocker. She couldn't tell if he was looking at them or the horizon or something else. He was very still.

"He looks...very well."

He did. *Better than before.* Not that she had a before, other than the specter-fying moments, but she had seen photos of him. Now he looked like a re-touched photo...

"Stay with the ship," Ernest called suddenly.

Prudence spun around. Winston and Fred were climbing over the gunwale. Ernest's order didn't stop them. She saw Niles and Crispin emerge from below and follow them off the ship. Out in the water, it almost seemed like the dolphin's fin changed shape. Franklin number four was on the move, too.

"Do you trust him?" Ernest asked, his hands tightening their grip on hers.

"Don't come any closer," Prudence called out.

Franklin four stopped. "You can't transit."

"Let's find out ."

"Wait." He moved to the place where the boardwalk from the cottage connected with that to the pier. He lifted a hand and the automatons stopped their advance.

Prudence flicked a glance toward Octavius. The dolphin—though now the fin looked like a shark's— was close, but not too close, she hoped.

"You're not my grandfather." Prudence said it, though she wasn't sure. His time on the ghost ship could have corrupted him. *Or he'd been corrupt when he disappeared.*

"You can run, but I'll find you," the possibly fake grandfather said.

"Not an answer," Ernest murmured. "I don't see your weapons?"

"Yeah, they didn't make it out with us here." Had they gotten out of something? Or fallen into something? When she'd been underwater, it had seemed very strange.

She eyed Ernest's ship. The automations were between them and the ship. Was that on purpose? Was it possible that the Blizablighter 3000 was back onboard? Was he right? Were they trapped? Or did they just want them to believe they were trapped? She hissed to Ernest, "See if you can keep him talking. Stall for time."

Ernest's nod was slight.

"Why do you care about us? Unless you seek revenge because we broke your ship."

The bubble edges quivered, as if Ernest had hit a mark.

"The ship was nothing. Only a small part of the plan." Franklin four smiled a mirthless smile. "I tell you again, you can flee, but you cannot hide from Him. There is no safe place."

"The Master?" Ernest asked. He looked around. "I certainly fear the *master* of this place." His tone was scornful, dismissive.

Again the edges quivered.

"Good job. Make him mad, hopefully, make him stupid." There was damage. Lines were broken and some were missing. It made the map unfamiliar, strange. She tried to remember, tried to find the line home—*there is no safe place.* What if home wasn't safe either? With The Storm doing so much damage, with so much danger, in hindsight, it seemed oddly easy that she got approval to come. Was there a traitor in the Agency?

"If the ghost ship was nothing, why have it at all?"

Franklin four chuckled. "Fear is so useful. And He needed it, he needs the lives it carried."

That seemed a telling slip. Had they freed his captives?

"It is obvious why He wants *her*, but you, you were a little mistake. You shouldn't have taken the stone. And now it is lost again." Franklin four sighed as if he regretted what must be. "And so you must be punished. If you step back from the human, Prudence, then his punishment will be...shorter. Otherwise, I'm very much afraid it will be eternal."

It was then the automatons rushed them, moving much faster than they should have been able to. And out in the water, the dolphin—no, the shark—surged toward Octavius, its suddenly huge mouth gaping wide.

"Hang on," Prudence said. Just before she closed her eyes, she saw Franklin four's face twist in rage, his visage flickering like a bad film for an instant. She thought she saw another face behind his, like an echo of a visage, but there was no time. She gathered Ernest and Octavius, then reached out for ship, felt the weight of it almost stall them out. As if Ernest sensed it, he sent her help as she groped for a line that wasn't broken within the morass inside her head.

She found one. Grabbed it. And with everything she had and a lot of Ernest's everything, she launched them toward somewhere.

Toward some when.

FIRE AND WATER. Ernest knew they didn't belong together, but somehow they were intertwined in this place. He felt the singe of fire as they passed out of the bubble dimension. How did he know this? It was as if Prudence's thoughts intermingled, flowed in and out of his. After the fire came the cool of water. This was better, though he felt the drag of a great weight. Prudence was there, her octopus and somehow so was *The Weatherman.* Lights shot past them as if the thing that called itself Franklin Pinkerton reached in to pull them back.

Now came the wind. Her eyes were wide and fixed, and somehow he knew she needed to concentrate on steering them, felt the strain of it in the way she clung to him. He tried to give her his strength, his confidence, but he knew, again, he did not know how, that there was much risk in this transit. That his ship, his beloved *Weatherman*, was part of the problem.

Let it go, he tried to tell her.

But then they hit rough currents. Buffeted from every direction, he hung on with gritted teeth. He felt her gasp, air rushed past them and then they slammed into something and went into a spin that got tighter and faster until the deep, silent dark swallowed them up...

PRUDENCE WOKE to the soft sound of waves slapping against wood and with a huge headache. Her head hadn't hurt this bad since her first transit—

She stiffened. *Transit.* Her eyes popped open. Above her, a dirigible tugged at the mooring that secured it to the ship. Ship. Airship. Really big airship. That would explain the rough, hard surface not helping her headache. And how tired she was. She was kind of amazed she'd got it here, wherever here was.

She should sit up, but in addition to the pain in her head, she felt like she'd been run through a wringer. Several times.

"Are you well, Prudence?"

Next to her Ernest eased himself into sitting position. It looked like he hurt as much as she did. She struggled into same the position and gave him an anxious once over. No blood or missing parts.

"I expect I will be well soon," she said. "You?"

"I am feeling rather battered," he confessed.

"That's because you were. Battered I mean. Dimensional transit leaves bruises." She flexed her legs. Found they worked, though they weren't happy about it. She was happy to have Ernest's help in getting to her feet, though it felt like they propped each other up in the process. "Oh man, that was a rough one." She arched her back, clenching her hands in the small of said back to ease the pain. It didn't work. She looked around. "I've never transited a ship to another dimension before. Did all of the *Weatherman* make it?"

She got a rather wide-eyed look from Ernest. His mouth opened, then closed. She executed a rueful shrug in answer to the question he didn't ask. Hey, from the outside he appeared he had the parts he needed. She must have looked that question because he nodded a bit dubiously.

"How about you check the ship and I'll—" her breath caught a little. If Octavius hadn't made it...

There was splash off the port side, and a tentacle appeared over the gunwale, rocking the ship slightly. She started to run—but scrapped that idea in a hurry and settled for hobbling over to the side.

"Are you all right, baby boy?" His dome rose out of the water, so she could stroke him. His color was normal. "Got all eight?" He drifted back and waved one at a time so she could count. Ernest appeared her side, giving Octavius a dubious look.

"It's all here." He cleared his throat. "*The Weatherman* is all...it is complete. I am rather amazed."

"I am, too, actually." Octavius bumped her hand with a tentacle, then indicated Ernest. "Sorry, baby. This is Ernest. Ernest, this Octavius."

With a stiff smile, Ernest took the extended tentacle. Another one snaked past and rubbed his head, then Octavius pulled back, sped away from the floating airship.

"He's probably hungry. I know I am." She looked at Ernest. "Do you suppose lunch made it here with us?"

His face shadowed, and she covered his hand with hers.

"I'm sorry about your automatons. I think he must have overridden their programming."

"Yes..." he frowned. "But I wonder if it was more than programming. They were more adept than they should have been."

"Really? I thought maybe you'd souped them up." He gave an inquiring look, so she added, "Made them work better."

"Souped up." He repeated the words, as if learning them, then asked, "Did you see Lord Henry and," he paused, then continued reluctantly, "Ophelia when they were—they had difference engine brains. Hybrid human and automaton, or so it seemed."

"Were they? I must have missed them going cyborg on us." Prudence looked at him anxiously. How did he feel about losing Ophelia again?

"Better," he said, slowly, "*better than before*. That's what he said."

"I do remember him saying that," she said. "Is that how you knew...?"

He nodded, his gaze distant as if remembering. "She, they thought I loved her, but it was because I realized I did not, that I never had loved her as much as she deserved, that I would have—but thankfully your plan worked."

She thought about several things she could have said, but cheering was a little insensitive, and she was too sore to cartwheel. Thankfully she had a thought.

"Do you think the automatons were hybrids, too, then?" Had he given them human brains or something? Ugh. None of it made sense unless she factored in the fact that somehow the—her mind flinched at using the name, but she didn't have another—Master could manipulate time. Or appear to manipulate it? At the very least, he or she had somehow constructed that bubble thing they were in, had scooped them out of the dimension they'd been in. Or just grabbed this small part of the dimension?

Real? Not real? Was it an illusion or a mix of reality and illusion? The Master had great power and ability and yet he had needed her grandfather, and then her...for what? Had the Master been after the Talent or was there some other purpose for capturing them?

"Possibly, though, I wonder, do you recall what he said about life forces?" His frown deepened and was tinged with worry. "Perhaps he knows how to...animate the automatons with the life force of the dead or recently dead? There were reports of people disappearing when the storm passed over."

"Like Frankenstein?" Prudence shuddered. Creepy enough to wonder if she would have been given one of the difference engine brains when it got through tormenting her but sending her life force into an automaton? That definitely boosted the creepy factor. "There did seem to be some sort of two-way thing going on there." She shuddered again. "I'm going to have nightmares for a long while."

"Me as well," he admitted. He seemed to give himself a slight shake. Then he smiled at her, and it was totally the jets where smiles were concerned. She had to smile back because she had a notion she didn't look the jets after the hammering and the drenching. The look in his eyes, a look she'd never expected or hoped to see *for her*, it took her breath away. It was amazing. It was entirely possible to be worried about the future of humanity and so happy her toes curled.

"So this is your *Weatherman?*" she said, not looking away because she couldn't. "You promised me a tour, I think." She was pretty sure her eyes weren't saying, give me a tour, but they probably should since they were alone on his boat somewhere and some when.

"I did," he agreed. He rubbed the back of his neck, then winced. His chest rose and fell, and his smile lost some of the heat. Though his gaze had a nice simmer in it.

Prudence drew in a shaky breath. To give herself a recovery

minute, she looked around, as in around past the airship. There was not a lot to see. Just water.

"I wonder where we are?" she murmured. For the first time, she let her brain wrap around the fact that she didn't know where they were or *where* they were. Or much of anything. Felt a bit like stepping onto that creepy gangplank. She felt her hand taken in a strong grasp.

"Let's take the tour, and you can tell me what is troubling you, my dear one."

Oh wow, that was brain-melting sweet. She looked at him and felt the worry melt a little. He led her toward the hatch and helped her down the short flight of stairs. In the narrow passageway, he directed her into a cabin that was obviously his research lab, or whatever he liked to call it. Probably a good thing he didn't show her his sleeping quarters. Not as tired as they both were.

His instruments were lined up on a wooden table built into the wall. They all appeared to be working normally. She looked at him, and he nodded. He pulled out his watch. Didn't know if it was the right time, but at least it wasn't going fast.

"Is the BlizaBlighter 3000 here, too?"

His eyes widened, and he quickly pulled open cabinet doors. His shoulders relaxed. "It is." He stepped back so she could see it, complete with the meteorite.

"He said it was lost," she murmured, touching it with the tip of her finger. No tingle this time. It was tapped out. She knew how it felt. Had she sensed it might be here? She'd grabbed it because it seemed the creepy guy didn't want her to have it, but maybe there'd been some instinct in there, too.

"What troubles you?" he asked again, his hands coming to rest on her shoulders.

"I don't know where we are," she admitted.

"It is a worry," he said, turning her to face him, "but once we see stars—"

"I mean, dimensionally."

"Oh. At least, I think I understand."

"I told you I can travel between dimensions. It's because I have something called the Talent. It allows me to transit dimensions, but the Talent, and with Octavius, well, with us working together we somehow find our way home."

He considered this, his head tilted to one side. "Without him, you could not?" She shook her head. "But you are no longer sure you can? Or you don't wish to go back?"

"A little of both," she admitted. "There are good reasons to go back and good reasons not to."

"What are the reasons not to go home?" he asked.

"Well, there's the meteorite that's not lost and the fact that I wasn't supposed to tell you anything about me or us or dimensional travel."

"Oh." He appeared about to ask, then just said, "oh," again. "And what are the reasons to go home?"

"Well, to report what happened. There's obviously something bad happening, that has been happening, only—"

"Only," he prompted when she stopped.

"Well, he said we couldn't hide anywhere. I wonder if there's someone inside the Agency," or more than one, she added reluctantly to herself, "and then there's the fact I'm not sure I can get home again." She added this lightly, hoping to soften the panic threading into her voice.

"But Octavius seems to be..." He stopped when she shook her head.

"The thing is, I see...transit lines with my mind. That's how I move around. Octavius stabilizes me somehow—the scientists that study familiars wonder if an animals' instincts play a part—but it's still about the lines. It's knowing what line to follow in relation to the others." She rubbed the sides of her head as if that would fix the problem. "Knowing what line and sticking it. That's the Talent."

"Something happened."

She nodded. "When my grandfather screamed, it...damaged my mental map. It was like a hammer to a mirror. Shattered lines, big gaps. And my tech all got fried." She tapped her arm, the one she'd used to create the sound-dampening field. "I hoped it was an illusion or temporary. But when I pulled us out of that bubble thing, I could see there was a lot of damage still. This line was the only one I could find that was complete. I mean, I think there are more that are okay, but I didn't have time— I had to go. But with the map damaged and my tech all offline, I don't know where we are. I don't know when we are. I'm not sure I can transit again or even if I should try."

"So we might be stranded here is what you are saying?"

He didn't sound too worried. He pulled her into a hug. "We are alive. We are together. We will figure out this place." He lifted her chin so he could smile into her eyes. "It is not so very bad having to live one life in one place—if you are living it with the one you love."

She felt her tension easing, a smile forming on her mouth once more.

"We don't know anything about this world. It could get interesting." She wasn't sure she was dressed right for this place. This made her want to laugh because it was such a girl thing to think. Her guidance counselor would be so proud.

"We will cross all bridges as we come to them." He bent his head, his lips pressing to hers for long enough to erase most of her worries. He lifted his head. "Things always look different when you are hungry. Let's see if there is something in the galley."

"And if not?"

"Then we will fish for it."

"Octavius can help with that."

"Already I see the benefit of having an, er, pet giant octopus."

She held him back with one hand. "And then?"

"And then we will figure out where we are in this world."

There was something in his voice that made her stop and look at him with arched brows. "And?"

"And how and where a man marries," he said, gathering into his arms, his voice husky as he added, "his one true love."

Prudence slid her arms up around his neck and pulled him close enough for kissing. "Oh, we are *so* going to find out how to do that..."

THANK you for reading *Specters in the Storm.* I hope you enjoyed it. There's an excerpt from my next book if you keep scrolling. :-)

To find out about all my releases, be sure to sign up for my New Release eZine and get a free eBook. You can also find out more by visiting my website.

If you enjoyed this book, I hope you'll consider leaving a review. It's not just because I'm needy (even though I try not to be!). Reviews help other readers decide which books to buy. :-)

SCYTHEQ: THE CYBORG CHRONICLES 6 SAMPLE

Introduction

Sparks fly between a half-cybernetic warrior and the no-nonsense pilot...

ScytheQ, once a fierce warrior now living in a sleek half-cybernetic body, has spent years battling her own emotions. She was forced to lock away her humanity behind the circuits and steel that she had become. When she is downloaded into a cloned body, her return to a human form leaves her yearning for connection.

Major 'Tiger' Teeger, a no-nonsense pilot still dealing with the aftereffects of a previous mission, has no time to deal with the mysterious woman on his ship, especially when sparks fly—both literally and figuratively—between them. All thoughts of ignoring Scy are shot to the outer limits of the galaxy when they encounter enemies lurking in the shadows and betrayal seems to wait around every corner. Trust becomes as deadly a weapon as any blaster in the remote regions of the star system.

As Scy grapples with her growing humanity and her blossoming feelings for Tiger, her feelings put them both in danger. The only way to survive is to face the truth about who she has become and the human she needs to be for herself and for Tiger, but will it be enough to save them?

Travel to the far reaches of the universe and discover the power of love and the human spirit as Scy and Tiger search for a safe harbor for their hearts.

Chapter One

SCYTHEQ FLEXED the cybernetic fingers of her left arm and eyed the looming figure of the OtimtronW watching her from the center of the exercise room.

Through the speakers, a song howled about werewolves, whatever they were. If Scy hadn't needed to keep on her toes, she'd have tapped them to the song. It was catchy. She made a mental note to find out what a werewolf was.

The OtimtronW robotic model was the most intimidating of the cybernetic models created by the Quy—and that was saying something.

All the models were big, dark, heavily armored and armed, and seemingly indestructible, but Tim was the biggest and darkest of them all.

Looks weren't everything. The cybernetic models had a weak spot—as the Quy found out a little too late to save themselves from the specially designed virus that had been unleashed on them. But that was long after their captain, CabeX, had exploited the cybernetics' internal systems and figured out a way to download their human consciousnesses into eight of the cyborg models. They'd had to leave their human bodies behind, but at least they'd been free and a version of alive.

They took the cybernetic ship they had been assigned, a ship designed for use by the cyborgs. And once freed, they'd become known as the most dangerous pirates in several galaxies, despite a bad habit of trying to defend the weak and free any slaves they encountered.

They'd have probably done more, but being relentlessly hunted by their former master—and the bounty hunters he'd hired—kept them on the move.

In their search for tech to keep ahead of their enemies, they'd arrived in the Garradian Galaxy and found unexpected allies and a way to become themselves again.

Scy had been one of the first to return to human form, not by choice but by necessity. It had been challenging, though, to be completely human after being so powerful. Fortunately, once that crisis was over, the Garradians had been willing to help her reintegrate some cybernetics back into her human body.

And Tim—the OtimtronW across from her—had agreed to let her try to pound on him to test that integration. He was a good friend.

"Are you ready?" Scy asked, mostly to stall as she shifted position, preparing to launch herself against all of that tall and unyielding metal. Maybe this wasn't such a good idea. The music changed to something about falling to pieces. She hoped it wasn't an omen.

Tim gave the cyborg equivalent of a sigh and crooked a metal finger at her. If he could have looked worried, he wouldn't have. There was literally no way her mostly human body could hurt him. But Tim wouldn't hurt her. Despite his imposing appearance, he'd always been a gentle soul.

The red glow of his optics deepened, as if mocking her for hesitating.

All right, then.

Scy ran forward, using enhancements integrated into her legs to run up the front of his body. With a twist, she was on his shoulders, one leg hooking around his neck. It was like trying to squeeze rock, but her inner systems assessed the force ratio of her output for later study.

A hand closed around her other leg and plucked her off, holding her so that her dark ponytail just brushed the mat.

She twisted loose—because he let her—and tried again, rolling under the length of an arm and attempting to trip him.

She failed, but still impressed herself with the amount of force she could bring to bear.

His hand closed around her neck—she'd told him to do it when it felt right to the moment—and lifted her off the floor again.

An inner shield blossomed out of her skin, keeping his hand from contacting her skin. It withstood the pressure he carefully brought to bear.

There was the sound of a throat being cleared and Tim swung around, taking Scy with him, her legs flailing.

Riina Katala stood in the doorway to the exercise room, her arms crossed and both her brows arched.

"Riina."

Cyborgs had little emotional range to bring to their voices unless they used a program. But Scy knew Tim liked Riina a lot. Everyone knew it except for Tim, as far as Scy could tell. And maybe Riina. She might not know. She wasn't a cyborg, but she was very good at hiding how she felt.

"Tim. Scy." Riina's voice gave little away, but the edge of her lips twitched. "Don't you think you should put Scy down?"

"Oh. Right."

Tim's hand opened and Scy flipped around and dropped lightly onto the mat, her hidden cybernetics easily absorbing the contact. She should have Tim throw her up and see how far the cushioning worked. But she wouldn't ask in front of Riina. Tim became less attentive when the Garradian was around—unless there was danger. Then he became hyper-focused.

The only two of their crew that had been human longer than Scy were CabeX and Boom. Both had successfully navigated the frontier of human emotions, or so it seemed, as they were both in relationships. Others of the crew who had become human after her had managed it as well. She supposed it gave her hope that she'd do it, too, though she still wasn't sure why she'd want a relationship.

She glanced around the well-equipped gym that she'd been using to test her physical limits and wished there was a gym for trying out human emotions.

She might have expressed this wish to CabeX during a weak moment. The transition back to a partial cyborg hadn't been without painful challenges. He'd smiled and shaken his head.

"You have to slug it out with love in the real world, Scy."

Maybe this love emotion was afraid to take her on?

She'd read some books supplied to her by women from the Earth Expedition, books they called "romances." The emotions the characters experienced had not yet affected her, though she'd paid close attention when she'd come within smelling distance or made eye contact with unattached males.

She wasn't sure if it was a plus or minus that the accumulation of males had gone down since CabeX had moved them to the Central Command of the Garradians. The Earth Expedition—and possessor of the highest concentration of eligible men—still preferred the Kikk Outpost, with its cloaking capabilities as their headquarters.

Travel between the two planets wasn't hard with the Garradian transit system, but she could hardly ask if she could go there to sniff men.

And considering the way the doctors had questioned her desire to reintegrate cybernetics into her body, she had a feeling that the changes might be off-putting to human males.

She'd gone ahead with the integration because feeling safe and in control was better than the male smells that might make her heart flutter. And that sounded very uncomfortable, anyway.

"How did it go?" Riina asked, her gaze moving between Scy and Tim.

When Scy had first transferred into this body, she had observed startled or uneasy glances from the other Garradians. Apparently, the clone source material for her body had been well-known to the other Garradians. But as her own consciousness had settled more deeply into the body, her face had undergone enough subtle changes that it happened less frequently.

She had a feeling that the cyborg enhancements would complete the disconnect for them. And the enhancements were helping her to feel more at home in this body—or at least more safe.

"She did well," Tim said.

"Then she is ready?"

"I am ready," Scy said.

Riina nodded. "I'll let the Maestra know."

THIS WASN'T the first time that Major Nathan "Tiger" Teeger had paid a visit to Garradian Central. The uneasy prickling along the surface of his skin was a pointed reminder of that last time—as if he needed to be reminded what a Charlie Foxtrot that had been.

Not that he blamed anyone but himself. He'd chosen to travel where no one had gone before, to meet aliens and shoot them when necessary. He'd had to shoot a lot of them. He tried not to dwell on the reality that this never seemed to reduce the potential aliens hoping to shoot back at him he tried not to dwell on. And not all of them wanted to shoot at them. Some aliens wanted to shoot with them.

He'd gotten both a promotion and new nightmares out of that

last alien shoot-up. He gave a mental shake of his head, remembering the psyche eval that had followed. Not exactly helpful when the shrink did not know how to deal with what had happened. So they cleared him to return to duty, and here he was.

He hoped he was keeping his expectations low enough for what was incoming. General Halliwell wouldn't have summoned him here just for a friendly chat about the non-weather or who won the championship back on Earth that they hadn't been able to watch.

Small talk was an interesting level of difficult in outer space. The movies, the books, the shows, the newspapers were all out of date by the time they reached this galaxy. And the faces? It was a good thing they had a planet to hang out on when things got too close on board the ship.

With all the close quarters and battles, it was a wonder the shrink'd had room on her schedule for him.

The blue line he'd followed since exiting the Garradian transit system stopped abruptly at a hatch.

"You have arrived, Major Teeger," a voice spoke from a hidden speaker.

It didn't sound like Bangle, but Tiger knew that the sentient AI could change her voice whenever she wanted to.

The hatch slid back before he could touch the controls and he entered, the hatch closing so quickly behind him, he felt it brush against his backside.

Had he pissed off the AI?

He came to attention when he saw General Halliwell sitting at the head of a long table. There were multiple screens on three sides of the room, each showing sections of space unfamiliar to him.

Or were they?

He didn't frown, not while he was at attention.

"Sit down, Tiger," the General said.

The general being friendly was never a good sign.

Tiger recognized the others at the table with him.

Doc the terrifying.

Nivi, the Maestra of the Garradians.

And a cyborg.

Yeah, he should be worried.

So that is who they waited for. Scy studied the male who had first come to attention, then removed his cap and taken a seat next to General Halliwell.

His sturdy frame inspired some confidence, she decided. Or was it he projected confidence, and she accepted his valuation of himself?

There was strength there, in both movement and in his square, determined face. Both he and the general wore similar camouflage which fit them comfortably, while still giving her the impression of competence.

His posture was almost as good as a cyborg's and his light blue eyes were keen, but had also turned wary at the sight of her.

He didn't like her? Or didn't trust her?

Whichever, it was mutual. She had protested the need to include an officer from the Expedition into her proposed mission.

CabeX, her captain, had overruled her, and Tals, another crew mate, had suggested she think of it as a first date. This reminder that every crew member who had taken a mission with someone from Earth had returned with a romantic partner did not amuse her.

She'd been even less amused when she'd researched dates and dating.

"I don't believe you've met Major Teeger," the Maestra said.

The soldier half rose and gave her a nod.

"I am ScytheQ," she said, returning his nod with careful cool.

"Pleased to meet you, ma'am," he said.

He didn't sound pleased. If anything, he sounded and looked even more wary. Was she so scary—oh? Perhaps this was what the doctors had meant when she wished to have cyborg augments, and they'd said it might make the other humans uneasy.

She found she didn't mind. She had missed being scary.

There was a pause, then General Halliwell shifted in his seat and said, "You believe you have a solution to our mutual problem?"

Scy glanced between him and the Maestra, but it was the one called Doc that the Maestra looked at.

"I," Doc said, "have gone over the data and I like it."

Scy knew her expression didn't alter. She still felt something of a disconnect between her consciousness and this face. Crewmates Boom and Mic had both suggested that practice in front of a mirror helped, but she couldn't see the point. How she felt inside changed so rapidly, it was hard to keep up.

This Doc certainly didn't worry about having expressions.

The general nodded and leaned back. He looked somewhat relaxed, but the way his index finger tapped the tabletop made her wonder.

Maestra gestured to Scy. "Can you brief the others, Scy?"

"Of course."

She triggered the hologram display she'd prepared for the Maestra, a data composite of the area between this outpost and what was called Victor Quadrant by the Earth expedition.

The Garradians had other names for all of their regions of space, but for the sake of clarity, the Maestra had encouraged her to use methods and language familiar to those in the expedition.

The major leaned forward, a deep crease between his marked brows.

"Is that Victor Quadrant?"

"It is," Scy said. "Using the data collected from the *Vega*, and from outposts within range and still able to collect data, we have been able to learn more about the anomaly."

The major shifted his position in his seat, as if it had suddenly become uncomfortable.

"Is there a problem, Major?" Scy asked.

"I know as much as I want to about the anomaly."

She accessed the crew data from the *Vega*.

"Oh, you were on board."

"Yes. And we were lucky to get out alive."

"Very lucky," Scy agreed, turning her attention back to her hologram.

"Our best scientists have been studying that data," the Maestra said, "with your scientists, General, and they feel fairly certain that what we are seeing is a collapsing galaxy." She glanced at Doc. "It is very exciting to have this data."

"You should try being in it," the major said.

"But you say it's not a threat to us," Halliwell put in.

"Not precisely," Scy said. "It will eventually eat its way through this galaxy..."

"But not for millions of years," Doc broke in.

Scy knew this was how they counted time on their planet, Earth. They didn't seem to realize how imprecise it was to impose their time on another galaxy. Being both a processor and a consciousness while inhabiting her robotic shell, she had a sense of time, but she didn't know how to turn that into something that could be counted.

She noticed the major appeared somewhat relieved.

"So we're just studying it, out of curiosity?"

"It is still a very large transit hazard," Doc said, "as you found out the hard way, Major."

"We've been working on a way to both warn travelers away and provide fixed points for someone to escape if they did wander inside it," the Maestra said.

The major tipped his head to the side. "Like a lighthouse?"

There was a pause. Scy sensed the others were considering this suggestion for its merit.

"Same principle," Doc said.

Scy used their interchange to do a data search for lighthouses. It was similar, while being wildly different. Interesting. Perhaps it helped them to find context?

"Scy?" The Maestra turned back to her.

Scy changed the hologram to show a schematic of the device she'd designed. She thought the major looked a little disappointed. It was true her design looked nothing like a lighthouse on his planet.

If she had been aware...no, he had not designed his lighthouses for space.

"The pod," Scy said, ignoring the major, "is both a bundle of sensors and gear for tracking the changes in the anomaly, and it will also broadcast warn-off signals in all known languages."

General Halliwell frowned at her hologram. "This pod will warn off traffic? It's so small."

Scy did not understand why size mattered, but said, "The devices are also large enough to provide emergency aid for distressed ships."

She provided them with a schematic of the interior, and both men were quiet while they studied it.

"It's like a camper in space," the major finally said.

Scy blinked, wondering why he sounded pleased.

"If needed, one or two scientists could stay on board while updating and collecting data," the Maestra added. "Or repairing damage."

General Halliwell frowned at the display. "How can you possibly plan for emergencies?"

Scy understood his concern. "We can't," she said. "But the pods will provide a safe place to send out a distress beacon if repairs aren't possible."

"We have outfitted more of the emergency response craft that AzumC tested for us," the Maestra said. "But the main purpose of the pods is to prevent ships from getting into trouble."

The major made a sound that sounded skeptical to Scy.

"Yes, Major?"

"That's a big, untidy and dangerous place." He looked like he wanted to continue, but compressed his lips into a thin line.

"The known edges of the anomaly are difficult to map," Scy conceded. "But we believe the pods will help us see what the naked eye cannot."

"I know what I saw," the major began.

"You saw the disruptions caused by the collapse," the Maestra said. "Planets ripped apart, debris fields, solar storms." She shrugged, her expression wry. "You are wise to be wary."

"We will attempt to place a pod close to where the *Vega* first encountered problems," Scy said. "We will activate it, monitor the data collection, and then, based on what we learn there, we will place another pod as far as we can while still keeping contact between the two pods."

"Why not just place them?" Halliwell asked.

"We don't know what we don't know yet," Scy said. The readings from the *Vega* indicate localized disruptions that will need to be accounted for.

"That sounds a lot like you're winging it," the major pointed out.

Scy blinked, but the Maestra chuckled.

"A concept you are most familiar with, I think?"

"She has us there, Major," the general said.

There was a pause, then Scy continued, "We are fortunate to have the data from the *Vega*. Using this known data..."

"Known data?" Doc avoided looking at her as she quietly offered the question. "Didn't you just say you don't know?"

Scy almost smiled. "Yes, there is much we don't know, but what we think we know," she shrugged. "The only way to map the unknown is to use the known."

Scy glanced at the general, then at the major. Both looked concerned. While she didn't blame them for this, she felt a tinge of impatience. She didn't need their permission to proceed with the mission. She wasn't sure why the Maestra wanted them "looped in" was the term she'd used.

Or why CabeX seemed to desire this as well. It was much too serious and, yes, dangerous, for his reason to be a match-making endeavor.

She saw the general hesitate and glance at the major. Was he also puzzled by the invitation to be part of the mission?

It felt as if the Maestra read Scy's thoughts when she said, "Together we see more and farther."

The general looked up, then his lips curved, though Scy wouldn't have called it a smile.

"Yes," he said. He once more looked at the major. He didn't speak.

"Yes," the major said.

Scy was fairly sure he sighed. That she understood.

GET *SCYTHEQ: The Cyborg Chronicles 6!*

Acknowledgments

I'd like to particularly thank Cornelia Amiri for assisting in the compilation of this little book on lighthouses.

And thanks to Bob for his assistance as well.

And as always, deepest thanks to my editor, Amy Brantley.

Bibliography

CHAPTER ONE: ANCIENT HISTORY TO THE RENAISSANCE

The Ultimate Book of Lighthouses: An Illustrated Companion to the History, Design, and Lore by Michael J. Rhein and Samuel Willard Compton. Published by Chartwell Books; updated edition (February 27, 2018).

Lighthouse of Alexandria, Wikipedia The Free Encyclopedia, https://en.wikipedia.org/wiki/Lighthouse_of_Alexandria

The Tower of Hercules: Two Thousand Years of Light
https://www.coruna.gal/the-tower/en/discover-the-tower/two-thousand-years-of-light/ancient-times?argIdioma=en#:~:text=The%20lighthouse%20of%20Ostia%20was,in%20the%20area%20of%20Ostia

Hercules Lighthouse: Everything You Need To Know, Attica21,
https://www.attica21hotels.com/en/the-tower-of-hercules/#:~:text=The%20Tower%20of%20Hercules%20is%20shrouded%20in%20legend.,Tower%2C%20associated%20with%20Irish%20mythology.

Lighthouses Of America, Book, by Tom Beard, Editor In Chief, Tom Thompson, Graphics Editor, published In 2017, by Welcome Books, a division of Rizzoli New York

CHAPTER TWO: HEROIC LIGHTHOUSE KEEPERS

The Complete Story of the Galveston Horror, by Various Authors, Book Online—Project Gutenberg's https://www.gutenberg.org/files/34304/34304-h/34304-h.htm

Kathleen Moore, 1812-1899 by: Mary K. Witkowski
https://bportlibrary.org/hc/maritime/kathleen-moore-1812-1899/

Kathleen "Kate" Moore, One of the First US Women Lighthouse keepers,
Women in Transportation History:
https://transportationhistory.org/2017/03/20/women-in-transportation-history-kathleen-kate-moore-one-of-the-first-us-women-lighthousekeepers/

Lime Rock Lighthouse *(Ida Lewis Lighthouse)* Rhode Island Lighthouses: http://www.rhodeislandlighthousehistory.info/lime_rock_lighthouse.html#anchorten

Three Black Lighthouse Keepers and Their Stories https://www.lighthousekeeperscottage.co.uk/black-lightkeepers-and-their-stories/

Chesapeake Chapter United States Lighthouse Society
https://cheslights.org/keeper-bio-venus-parker/
and https://cheslights.org/keeper-bio-william-major-parker/

William Major Parker: The Pioneering Black Keeper at Killick Shoal Lighthouse
https://www.secretsoftheeasternshore.com/william-parker-killick-shoal/

The Mingo Chronicles by Thomas Tag https://uslhs.org/sites/default/files/articles_pdf/BlackLighthouseKeepers_TheMingoChronicles_Summer_2009.pdf

Makah Tribe history tells us Makah life revolved around the sea and land, About our History
https://makah.com/makah-tribal-info/about-history/

Cape Flattery Light on Tatoosh Island begins operating on December 28, 1857, HistoryLink.Org Esssay #5703 by William S. Hanable, posted 6/08/2004
https://www.historylink.org/File/5703

Keeper Bio: William Roscoe Davis, Chesapeake Chapter United States Lighthouse Society
https://cheslights.org/keeper-bio-william-roscoe-davis/

Underwater Panther, Wikipedia, the free encyclopedia—
https://en.wikipedia.org/wiki/Underwater_panther

Menorcan Keepers at the Old St. Augustine Lighthouse
The Old Spanish Watchtower https://www.staugustinelighthouse.org/2016/03/23/menorcan-keepers-at-the-old-st-augustine-lighthouse/

CHAPTER THREE: LIGHTHOUSES DURING WARTIME

Lighthouses Of America, Book, by Tom Beard, Editor In Chief, Tom

Thompson, Graphics Editor, published In 2017, by Welcome Books, a division of Rizzoli New York

Scheveningen Lighthouse —September 2005 edition of Lighthouse Digest Magazine of edit of Lighthouse Digest®, Inc.

Gun Fight at the Cape Florida Lighthouse, Magic Masts and Sturdy Ships, https://magicmastsandsturdyships.weebly.com/gun-fight-at-the-cape-florida-light house.html

Point Lookout Prisoner of War Camp, The American Civil War, https://www.mycivilwar.com/pow/md-point-lookout.html

From Gray to Blue, Civil War Era, Muster How The Past Informs The Present: An Odyssey of Deserting the Confederate Army and Joining the U.S. Army, *By Kyle Nappi | June 15, 2021*
https://www.journalofthecivilwarera.org/2021/06/from-gray-to-blue-an-odyssey-of-deserting-the-confederate-army-and-joining-the-u-s-army/

Lighthouses In The Civil War, Civil War Talk
https://civilwartalk.com/threads/lighthouses-in-the-civil-war.118079/

Lighthouses in the UK during WWII, Lighthouse Stories and Writing Encouragement, Marilyn Turks's Pathways of the Heart
https://pathwayheart.com/lighthouses-in-the-uk-during-wwii/

Lighthouses In World War I: **Transition into War** by Ellen Henry, Curator, Ponce De Leon Inlet Lighthouse Preservation Association
https://media.defense.gov/2017/Aug/07/2001789112/-1/-1/0/2017-HENRYEL LENUSLHS-WWI-ARTICLE.PDF

CHAPTER FOUR: HAUNTED LIGHTHOUSES

Owls Head ME, Owls Head Lighthouse, 2001-2025 Lighthousefriends.com
https://www.lighthousefriends.com/light.asp?ID=527#:~:text=Owls%20Head%20-Lighthouse%20is%20number,or%20looking%20out%20a%20window.

An Old Haunted Lighthouse Was Voted The Creepiest Place In Florida, Author: Joe Winner
https://sunny1063.com/2023/10/30/an-old-haunted-lighthouse-was-voted-the-creepi est-place-in-florida/

Guardians Of The Lights : Book—The Men And Women Of The U.S. Lighthouse By Elinor De Wire, Pineapple Press; Reissue Edition (June 1, 1996)

CHAPTER FIVE: EXTRAORDINARY LIGHTHOUSE PETS

Guardians Of The Lights : Book—The Men And Women Of The U.S. Lighthouse
By Elinor De Wire, Pineapple Press; Reissue Edition (June 1, 1996)

SOURCES USED IN SEVERAL CHAPTERS

U. S. Coast Guard Historically Famous Lighthouses.PDF CG-232 https://media.defense.gov/2017/Jul/03/2001772530/-1/-1/0/H_FAMOUSLIGHTHOUSES.PDF

Book—Lighthouses Of The West Coast: Washington, Oregon, and California by Ray Jones, published in 2023 by Globe Pequot, the trade division of the Roman and Littlefield publishing group, ink.

Book: Brilliant Beacons: A History of The American Lighthouse, by Eric J Dolan, published in 2016 by Liveright; Illustrated edition (April 18, 2016)

Book: Lighthouses: A Photographic Tour, Captions by Julie Stetzko Taff, published in 1998 by Brown Trout Publishers Inc., San Francisco, California

Book: Lighthouses by S. Ross Holland, published in 1997 by Barnes and Noble Books, New York

Book: Lighthouse Encyclopedia: The Definitive Reference by Ray Jones, Globe Pequot (February 20, 2017)

Books by Pauline Baird Jones

Science Fiction Romance/Paranormal

Project Enterprise: The Cyborg Chronicles (Coming in 2023!)

Cyborg's Revenge: The Cyborg Chronicles Book 1
Cosmic Boom: The Cyborg Chronicles Book 2
CabeX: The Cyborg Chronicles Book 3
AzumC: The Cyborg Chronicles Book 4
MircoP: The Cyborg Chronicles Book 5
ScytheQ: The Cyborg Chronicles 6
TalusH: The Cyborg Chronicles 7
OmnitronW: The Cyborg Chronicles 8

Project Universe Series:

The Key (book 1)
Girl Gone Nova (book 2)
Tangled in Time (book 3)
Steamrolled (book 4)
Kicking Ashe (book 5)
The Reboot Books of Project Enterprise
Found Girl (book 6)

Lost Valyr (book 7)
Maestra Rising (book 8)
More Project Enterprise
Project Enterprise: The Short Stories
Time Trap: A Project Enterprise Series Short Story
Operation Ark: A Project Enterprise Story
General's Holiday: A Project Enterprise Story
Claws & Effect: The Otherworldly Pets of Project Enterprise

Other Romantic Science Fiction Stories
The Real Dragon
Nebula Nine (time travel adventure)
Open With Care (Christmas collection that includes, "Riding For Christmas" and "Up on the House Top"
Specters in the Storm: A paranormal/steampunk/science fiction romance novella

Out of Time Series:
Out of Time (World War II Time Travel Romance)
Just in Time (An Out of Time Story)
Telling Time (January 2024)

An Uneasy Future
(A science fiction romance mystery series set in future New Orleans)
Core Punch (1.0)
Sucker Punch (2.0)
One Two Punch: An Uneasy Future Bundle

Romantic Suspense

The Big Uneasy Series:
Relatively Risky (1)
Family Treed (A Big Uneasy Short Story)

Dead Spaces (2.0)
Louisiana Lagniappe (3.0)
Worry Beads (4.0)
Fais Do Do Die (5.0)
Beaucoup Fracas (6.0)
Pirogue Wipe Out (7.0)
Bourre Brouhaha (8.0)
The Family Way (A Big Uneasy Short Story)
Guess Who's Coming To Christmas: The Wedding Edition
The Big Uneasy Bundle
An Uneasy Collection: The Big Uneasy Books 3-5

Lonesome Lawmen Series:
The Last Enemy
Byte Me
Missing You
Lonesome Mama (Bonus short story)
(The *Lonesome Lawmen* is also available as a digital bundle)

Do Wah Diddy Die
The Spy Who Kissed Me
Perilously Fun Fiction Bundle (includes *The Spy Who Kissed Me* and *Do Wah Diddy Die.* Bonus: *Do Wah Diddy Delete Short Story Collection)*
Dangerous Dance
Dangerous Duet

Short Story Collections

Project Enterprise: The Short Stories
Do Wah Diddy Delete
Let's Fall in Love
The Real Dragon and other short stories

Non-Fiction

The Author's Odyssey: Humorous & Real-world Insights from a 25-year Journey in Publishing

Penning Pulse-Pounders: A Spirited Guide to Writing Your Own Suspense Novel

Guiding Lights: A Tribute to Earth's Lighthouses

About the Author

Award-winning, *USA Today* Bestselling author, Pauline never liked reality, so she writes books. She likes to wander among the genres, rampaging like Godzilla, because she does love peril mixed in her romance.

To find out more about Pauline or her books:
http://paulinebjones.com
pauline@paulinebjones.com

Made in the USA
Monee, IL
04 November 2026

67385795R10118